Elevator Design, Construction & Maintenance

- 1905 -

Elevator Design, Construction & Maintenance

- 1905 -

Compiled By

Copyright © 2008 Merchant Books

ISBN 1-60386-116-5

DIGITALIZED BY
WATCHMAKER PUBLISHING
ALL RIGHTS RESERVED

ELEVATORS.

(PART 1.)

GENERAL DESCRIPTION OF ELEVATORS.

INTRODUCTION.

1. Definition.—The term **elevator** is applied to that class of hoisting machinery in which a cage, cab, car, or platform is raised and lowered between fixed stops or landings.

2. Principal Parts.—In all complete elevators the following principal parts are easily distinguished:

1. The motor.
2. The car (cage, cab, or platform) and its principal guides.
3. The devices transmitting power from the motor to the car.
4. The counterbalance weights and their guides.
5. The controlling devices.
6. The safety devices.
7. Accessories.

MOTORS AND CLASSIFICATION.

3. Various kinds of motive power and, consequently, motors are used to run elevators. In practice, the classification of elevators is made according to the motive power used. The most generally accepted one, which is also the one that we shall adopt, is as follows: *Hand-power elevators*, *belt elevators*, *steam elevators*, *electric elevators*, and *hydraulic elevators*.

CARS AND GUIDES.

4. It is evident that elevator cars must be different for various purposes. All of them, however, have a **platform** upon which the load rests, and with few exceptions, as in sidewalk elevators, two upright **posts** connected by a **crosshead** to which the ropes are attached. Each upright carries two **guide shoes,** one on top and one on the bottom, which fit over the **guides.** The latter consist either of hardwood strips of square cross-section or bars of **T** iron carried up inside the hoistway and attached to suitable supports. According to the location of the elevator shaft in the building and the accessibility of the guides, they are placed either in the center of two opposite sides of the shaft or in two diametrically opposite corners, necessitating the upright posts of the cars to be placed in like manner with reference to the platforms. In the first case they are called **side-post elevators;** in the other case, **corner-post elevators.** The guide shoes are usually of cast iron, and in the case of iron guides are lined with Babbitt metal.

For freight service the cars are of the simplest kind; they are generally made of wood with iron fixtures and bracings. For passenger service a complete cage is built upon the platform, preventing any possible contact of the passenger with the hoistway. Passenger cars are now mostly built wholly of metal, though many wooden ones are in operation. Various styles of cars are shown in subsequent illustrations.

TRANSMITTING DEVICES.

5. Various transmitting devices are used with different kinds of motive power. Hydraulic elevators and a certain electric elevator have peculiar transmitting devices of their own, which will be described in connection with these elevators. All belt and steam elevators and the majority of hand and electric elevators are of the **drum type,** that is, of a type in which the transmitting devices include a drum and rope. All these elevators, therefore, have certain peculiarities in common, which are pointed out beforehand to avoid repetition.

6. Side Travel of Ropes in Drum Elevators.—An inherent feature of the rope-and-drum drive is the deflection of the rope as it winds upon the drum. Let D, Fig. 1, be the winding drum and S the nearest sheave from which the

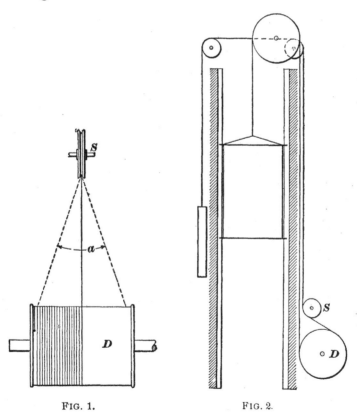

FIG. 1.　　　　　FIG. 2.

rope passes on to the car. It is plain that only at a certain position of the car the rope runs over the sheave exactly straight; in all other positions it must be guided into the sheave. If the distance between the drum and the nearest sheave is great, as, for instance, when the rope passes straight up from a drum located at the foot of the hoistway to an overhead sheave, the deflection measured by the angle a, Fig. 1, is but small and readily taken care of by the depth of the grooves in either sheave or drum. But if the distance is small, danger exists that the rope will jump the grooves of the drum and "ride" on itself, which may evidently cause accident. Such small distances between the drum and the nearest sheave are frequently unavoidable. Thus, in the case shown by Fig. 2, where it is required that the ropes of both the car and the counterweight shall run within the hoistway, the hoisting rope must be led over an **idler** S very near the drum D, and the counterweight rope, in the case shown, will surely "ride" if no provision be made against it. These idlers are, therefore, so mounted on their shafts that they can follow the ropes as they wind upon and unwind from the drums. Such a traveling idler is sometimes spoken of as a **vibrator**. In most cases it is found sufficient to mount the idler loosely on a smooth shaft and to rely on the pressure of the rope against the sides of the groove in the idler to shift the latter along. That careful lubrication is essential to the proper working of this arrangement is evident.

7. The constant chafing of the rope against the sides of the idler groove, which is unavoidable in the arrangement just mentioned, is an objection, and if considered of sufficient influence on the life of the rope, is avoided by giving the idler a positive motion in the direction of its shaft. Figs. 3 and 4 show two ways of accomplishing this.

In Fig. 3, the idler shaft a is connected to the drum shaft by a chain and sprocket wheels, the hub of the sprocket wheel b on the idler shaft being a nut fitting over a square thread cut on the shaft and being held from moving sidewise. This causes the shaft to move in the direction of its axis, a feather preventing it from rotating. The idler c turns loosely on the shaft, but moves back and forth with it, due to collars on the shaft.

8. In the arrangement shown in Fig. 4, the chain connection is dispensed with. The idler shaft a is threaded but is held stationary, and the idler hub is a nut, so that while the idler revolves by the friction of the rope it travels back and forth. Since there is no positive connection between the drum shaft and the idler shaft, any slippage of the rope on the idler will bring the arrangement out of adjustment. For this reason the following plan for automatic readjustment was adopted by the Otis Elevator Company.

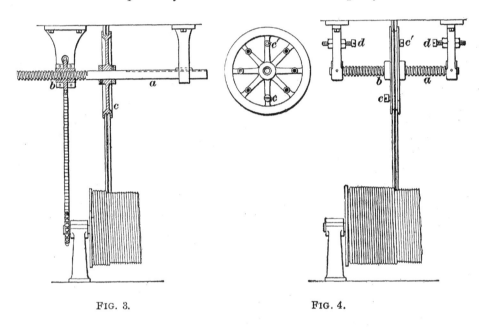

FIG. 3.   FIG. 4.

The idler is provided with stop screws c and c' that engage at the end of the travel with fixed stops d, d on the shaft supports. If for any reason slippage has occurred and the idler lags behind, it will be ahead of the rope on the following return trip and engage the stop on the shaft support before the drum comes to rest; the idler being thus prevented from turning, the rope will slip until the drum stops; on the following trip the idler will leave the stop at once and, thus readjusted, will follow the rope correctly. To allow of a fine initial adjustment, the idler has eight spokes, each drilled and tapped to receive the screw stops c, c'.

9. Absorption of Vibration Due to Gearing.—An inherent feature of all drum elevators is a certain amount of unpleasant vibration transmitted from the gearing through the drum and hoisting rope to the car. This vibration is especially noticeable in spur-geared machines; but it also exists in worm-geared ones, owing to the fact that a certain amount of backlash, be it ever so little, always exists. To reduce these vibrations to a minimum, elastic buffers, generally of rubber, are sometimes interposed between the drum and the next adjacent gear. Fig. 5 shows a way in

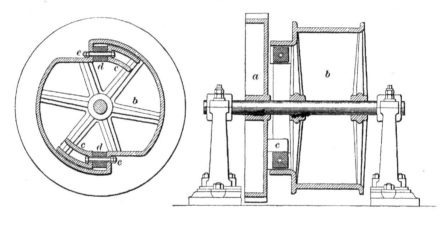

FIG. 5.

which this may be done. The gear-wheel a is keyed to the shaft, while the drum b is loose. The gear-wheel drives the drum by means of drivers c, c, which are cast in one with the gear-wheel. These drivers, instead of butting directly against metallic surfaces of the drum, butt against rubber blocks, or buffers, d, d. These buffers must be given a certain amount of initial tension, which is accomplished by the tie-bolts e, e that tie the drum and the gear-wheel together. The end view shown in Fig. 5 is taken between the gear-wheel a and the drum b, which accounts for the fact that while the drivers c, c are seen, the gear-wheel is absent. The tie-bolts must have jam nuts or some other efficient nut-locking device, which should be examined occasionally to see if the bolts have become slack.

COUNTERBALANCING.

10. In any elevator, the weight of the car and its fixtures is constant, and hence is easily **counterbalanced** to any extent desired. The simplest way to do this is to attach another rope, besides the hoisting rope, to the car, leading this second rope over one or more overhead sheaves and suspending from it the counterbalance weight, or the **counterweight,** as it is generally called, as shown in the diagrammatic illustration given in Fig. 6. In order that the car may descend when empty, the counterweight must, when so attached, always be less than the weight of the empty car with its fixtures. Evidently, with such an arrangement, no power is needed for the down trip of the car, while on the up trip, the motor must develop enough power to raise the maximum load, plus the unbalanced weight of the car. In all types of elevators in which the motor furnishes power only on the upward trip of the car, as in hydraulic elevators, for instance, the arrangement shown in Fig. 6 is the only method of counterbalancing available.

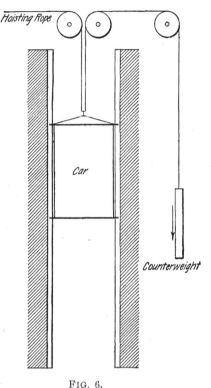

FIG. 6.

11. If in an elevator the power can be applied during the down trip as well as during the up trip, then not only the full weight of the car can be counterbalanced, but also a part of the load. An elevator thus counterbalanced is said to be **overbalanced.** This possibility exists in all drum elevators, as the motor and drum are reversible. They are, therefore, overbalanced, except when other considerations make it undesirable, to the extent of the average load by attaching the counterweight to the drum and winding the rope in an opposite direction to that of the hoisting rope, as shown in Fig. 7.

The advantage of overbalancing is easily apparent. If the load on the car is equal to the average load, no power is

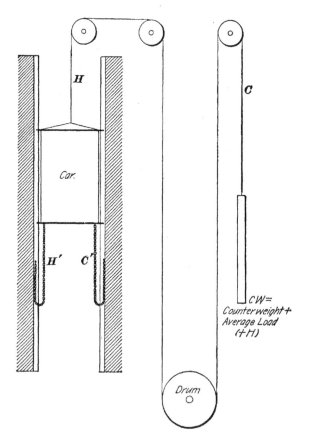

FIG. 7.

needed besides that necessary to start the machinery and keep it moving against frictional resistances. If the load is equal to the maximum load and the car is going up, the motor must furnish power enough to raise the difference between the maximum and the average load; or if the latter is one-half the maximum load, to raise one-half of the maximum load. If the car is going down empty, which is the other extreme possibility, the motor must raise the counterweight, that is, the weight of the average load. Thus a motor can be used of greatly smaller capacity, which means smaller size, less weight, and smaller cost. In connection with electric drum elevators, overbalancing also tends to equalize the current consumption.

12. By an arrangement somewhat different from that shown in Fig. 7, the stress in all the ropes and the pressure on the drum-shaft bearings may be diminished. In Fig. 8

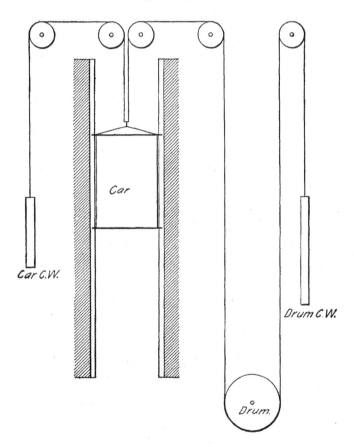

FIG. 8.

there are shown two counterweights, one attached directly to the car and the other to the drum. The car counterweight must evidently be less than the weight of the car in order to allow the car to descend when empty; the other counterweight is equal to the remaining portion of the car weight plus the average load. If the car counterweight is, for instance, one-half of the car weight, then the stress in the drum-counterweight rope and the hoisting rope is less than in the arrangement shown in Fig. 7 by one-half of the weight of the car, and the pressure on the drum-shaft bearings is less by the whole weight of the car.

13. For high lifts, the weight of the ropes themselves is a considerable item, making the counterbalancing change for different positions of the car. To avoid this, balancing chains having the same weight as the ropes to be balanced are used and are hung from the bottom of the car, either reaching all the way down to the bottom of the hoistway, in which case the chain must have the same weight per unit of length as the ropes, or reaching down only to the middle of the shaft and fastened there, in which case the chain must have a weight per unit of length double that of the ropes to be balanced. This method is indicated in Fig. 7.

The ropes to be balanced here are the hoisting rope from the car to the overhead sheave and the counterweight rope from the counterweight to the overhead sheave, denoted, in Fig. 7, by H and C, respectively. The former can be balanced by a chain H' of equal weight and an increase of the counterweight by the same amount, while the rope C can be balanced simply by a chain C'. Of course, two chains would be actually used, each weighing $\frac{H+C}{2}$.

14. All counterbalancing means an addition to the moving masses of the elevator, which, again, means an increase in the power required to set these masses in motion, as well as greater braking power to stop them. These considerations make it desirable in certain elevator types to forego the advantages of overbalancing.

COUNTERBALANCE WEIGHTS AND GUIDES.

15. The **counterweights** consist generally of cast-iron blocks carried in a frame or on a rod or rods and guided by suitable guideways. The blocks are made long and wide, but thin, in order that they may take up but little room. In hydraulic elevators the counterweights are sometimes attached to the piston rods, either inside or outside of the hydraulic cylinder.

The counterweight guides are made of angle or T iron, seldom of wood.

CONTROLLING DEVICES.

16. Kinds of Controlling Devices.—The controlling devices of all elevators consist of a **power control**, that is, means for shutting off, turning on, and regulating the power at will to start and stop the car, and some kind of a **brake**, the function of which is to effect a prompt but gradual, and therefore safe, stoppage of the car.

The power control and brake are essential parts of the motor in each case and are, therefore, located near the same; they are naturally different for different kinds of motive power, and will be described at length in connection with the various types of elevators. There is, however, with respect to the controlling devices a certain feature common to all.

Since most elevators are operated from the movable car, some flexible connection must exist between the same and the controlling devices on the motor. The means for making this connection, which we will call **operating devices**, are either mechanical or electrical. The latter is used to any extent only with a certain kind of electrical elevators, while the mechanical connection is employed on all types in the shape of a **shipper rope** running all the way from the top to the bottom of the hoistway and either simply passes through the car or is connected with some apparatus inside the car.

17. Different Operating Devices. — The simplest arrangement is a plain endless rope hung over one or more idlers and attached to the **shipper sheave** or a lever, so that a pull either up or down on the shipper rope moves the sheave or lever, which is located on or near the motor and is mechanically connected to the controlling devices of the same. This simple arrangement, which is shown diagrammatically in Fig. 9, is open to several objections, one of which makes its use undesirable in connection with all motors requiring a delicate adjustment of the controlling device, such as hydraulic motors controlled by a pilot valve and electric motors, inasmuch as the operator has no means of telling the exact position of the controlling device. Another objection is that there is necessarily a great deal of sliding of the rope through the hand of the operator, which is not only inconvenient, but may prove dangerous from broken strands. The operator should provide himself with a leather glove or use a piece of rubber hose split lengthwise.

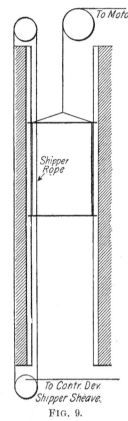

Fig. 9.

18. In order to overcome the objections to the simple shipper rope, various devices have been invented and put into use. They all have the object of changing the up-and-down pull of the rope into the motion of a lever or crank. In Figs. 10 to 16 are shown a number of these devices as actually used, particularly in connection with hydraulic elevators. In the arrangement shown in Fig. 10, there is a three-armed lever A in the car, the long arm of which is to be swung to the right or left by the operator. To each of the short arms is connected a rope R running down over an idler carried by another three-armed lever B pivoted at the bottom of the hoistway. From the idlers on lever B the ropes R, R pass up again and over idlers fixed at the top of the hoistway, as shown. On the ends of the ropes are counterweights which are equal, and each is somewhat heavier than the equivalent force necessary to move the controlling device of the motor, which is mechanically connected to the third arm of the lever B. As will be easily understood, the whole arrangement is in equilibrium in any position of the lever A, but the equilibrium is disturbed as soon as a pull is exerted on it in either direction, in consequence of which the second lever B will follow the motions of A and stay in a position corresponding to that of A. The top idlers are shown in the diagram on separate shafts or studs; in reality they are placed side by side and the downward

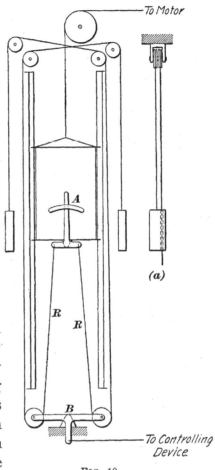

Fig. 10.

rope passed through the counterweight nearest to it, as shown in Fig. 10 (*a*). The weight is thus guided on the rope. To prevent abrasion of the rope, a rubber ring is inserted in the hole through the weight.

19. Another arrangement is shown in Fig. 11. The shipper rope is led from a fixed point *a* at the top of the hoistway over two idlers *b* and *c* mounted on a lever L pivoted to the car and handy to the operator. From the idlers *b* and *c* the rope is carried farther down, around the shipper sheave S, and thence back upwards over two more idlers c' and b', also attached to the lever L, and is finally fastened at the top at d.

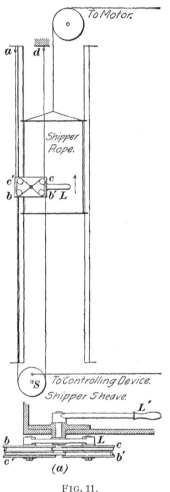

FIG. 11.

While the car moves up and down, the shipper sheave S is stationary unless the lever L is moved. By moving L upwards, the part $b\,c\,S$ of the rope is doubled up more, while the part $b'\,c'\,S$ is straightened out an equal amount, causing the sheave S to take a position depending on that of the lever L, in which position it remains until the lever is moved again. In an actual machine the idlers are mounted on studs, b and c' on one stud and b' and c on another stud, which are carried on a lever outside the car; the pivot of the lever is carried through into the interior of the car, where it carries the handle L' [see Fig. 11 (a)]. A hand wheel may be substituted for the handle L'.

20. The same idea that underlies the arrangement of Fig. 11 is embodied in Fig. 12. Here the two branches of the rope are deflected so as to pass over two idlers i, i' on the same stud and are attached to a lever pivoted to the car; the other idlers a, b, c, and d are fixed to the car.

21. The devices shown in Figs. 11 and 12 necessitate idlers to be carried on the car, where they must, owing to the limited space available, be necessarily small. This is detrimental to the rope, especially since it is bent in opposite directions in quick succession. These objections do not prevail in the arrangement shown in Fig. 10, where the idlers may be ample in diameter and the ropes are bent in one direction only.

22. An improvement on Fig. 12 is the operating device shown in Fig. 13, in which the ends of the ropes are attached to the car instead of to moving weights; a single

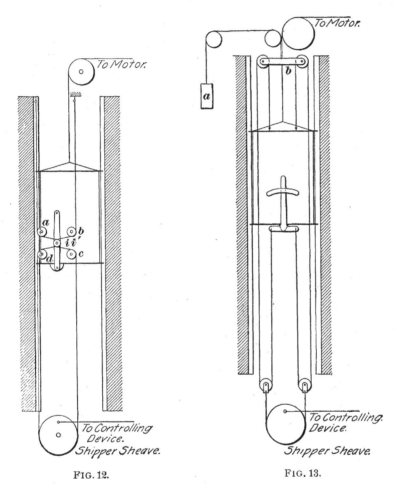

FIG. 12.   FIG. 13.

stationary weight *a* attached by a rope to a cross-bar, or frame, *b* carrying the upper idler is substituted for the moving weights.

23. This arrangement has been still further improved upon in the manner shown in Fig. 14. The ends of the ropes that were attached to the car in Fig. 13 are here also attached to the lever arms, and the two ropes leading from the lever of the two idlers are crossed. It can easily be seen that by this means the motion of the lever gives twice as much motion to the sheave as in the arrangements shown in Figs. 10 and 13.

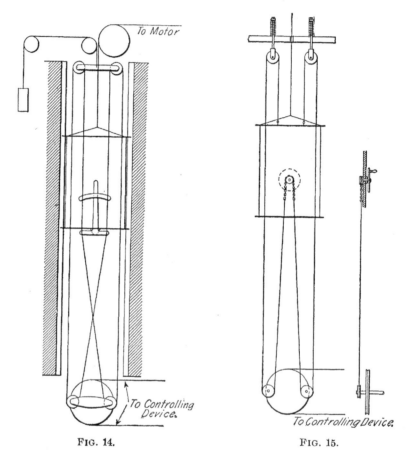

FIG. 14.   FIG. 15.

24. The operating device last described is known as the **Otis lever,** or **operating device,** and is now used almost exclusively on the hydraulic elevators of the Otis Elevator Company. By substituting a hand wheel or crank for the lever in Fig. 13 and attaching the lower idlers to the shipper sheave, a modification shown in Fig. 15 is obtained that is found on a good many elevators of the Otis make and is called a **hand-wheel operating device.** By elevator

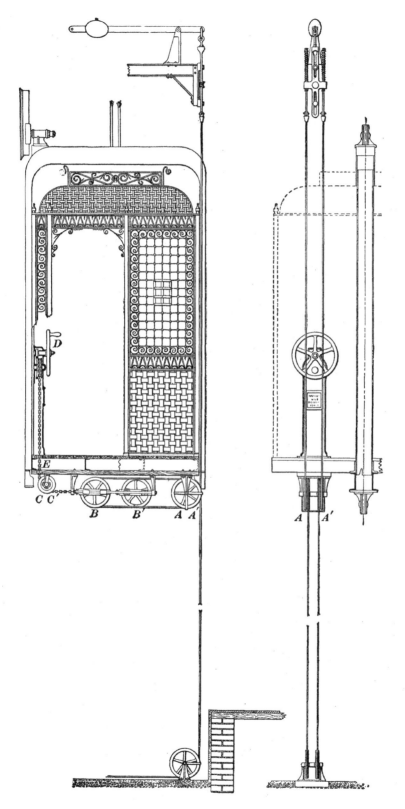

Figure 16

runners the operating devices are often called **controllers** and are spoken of as **lever controllers, hand-wheel controllers,** etc. Since the term *controller* is also given to the combination of switches and resistances constituting the controlling device proper in electric elevators, the practice just mentioned is not followed here in order to prevent confusion.

25. A common arrangement of a hand-wheel operating device is shown in Fig. 16. The sheaves A, A' are stationary, but the sheaves B, B', being loosely mounted on guide rods, can be shifted by means of the hand wheel D and chain E. The chain runs over the wheels C, C'. After the explanations of the operation of the different operating devices that have been previously given, the operation of this one will be easily understood.

SAFETY DEVICES.

26. We can divide the **safety devices** used on elevators into two distinct classes: those that control the power supply, which we shall call **motor safeties,** and those that control the car independently of the power supply, which we shall call **car safeties.** The former must necessarily be treated in connection with the various styles of motors used; the latter allow of, and their importance warrants, a treatment by themselves, which will be given in its proper place.

ACCESSORIES.

27. Among **accessories** we shall class all those various appliances designed to prevent accidents from causes other than failure of the elevator or any of its parts, and to increase the comfort of the traveling public and the efficiency of the elevator service. Such appliances are automatic gates, doors, and hatchways, signals, indicators, etc.

HAND-POWER ELEVATORS.

CONSTRUCTION.

28. When an elevator is to be used but little, and especially if speed of the car is not essential, it does not pay to use steam or other motive power; **hand-power elevators** are then useful. With few exceptions they are installed for freight service only.

Figs. 17, 18, 19, and 20 show several types of hand-power elevators. Those shown in Figs. 18 and 19 are made by Morse, Williams & Co., of Philadelphia, Pennsylvania, and those shown in Figs. 17 and 20 by the A. B. See Manufacturing Company, of Brooklyn, New York.

29. Motor.—The *motor* of a hand-power elevator is represented either by a shaft actuated through a rope sheave and endless rope, the latter being pulled in either direction by hand, and examples of which are shown in Figs. 17, 18, and 19, or it is represented by a crank driving a windlass, as shown in Fig. 20.

30. Transmitting Devices.—The *transmitting devices* consist of spur gearing in connection with either a drum for a rope or chain, as in Figs. 17, 19, and 20, or a friction sheave, as in Fig. 18.

31. Operating Devices.—The *operating device* is a manila rope, preferably a four-strand and "stevedore" rope; the hoisting and counterweight ropes are generally wire ropes. In the sidewalk elevator shown in Fig. 20, chains take the place of the rope.

32. Cars.—The cars in Figs. 19 and 20 are different from the ordinary cars, inasmuch as they are supported on all four corners.

Small hand-power elevators are used largely for carrying small loads in dwellings, restaurants, libraries, etc., and are called **dumbwaiters**. The cars of these then take the shape of a box with or without shelves.

33. Guides and Counterweight.—In the elevator shown in Fig. 19 *guides* are provided on one side only. The *counterweight* in Fig. 17 is hung from a separate drum;

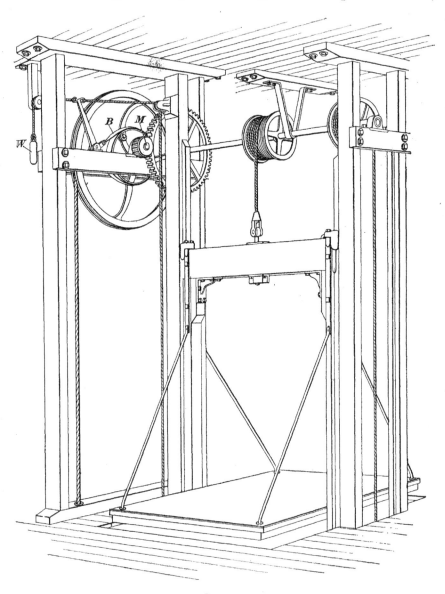

FIG. 17.

in Fig. 19 it is hung from one of the hoisting drums; in Fig. 18 it is attached to the other end of the hoisting cable; and in Fig. 20 the counterweight is dispensed with entirely.

When hand-power elevators are balanced, they are generally overbalanced.

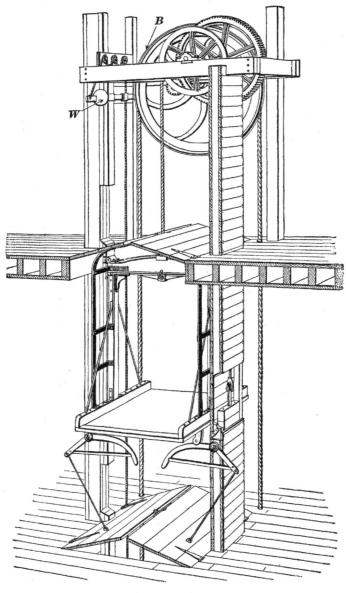

Fig. 18.

34. Controlling Devices.—The *controlling device* in Figs. 17, 18, and 19 consists only of a brake *B*, which is applied by a weight *W* and is loosened by the operator by means of a hand rope. In the windlass, or winch, type of elevator, shown in Fig. 20, the brake is actuated, both in applying and loosening it, by operating the lever L by hand. Since

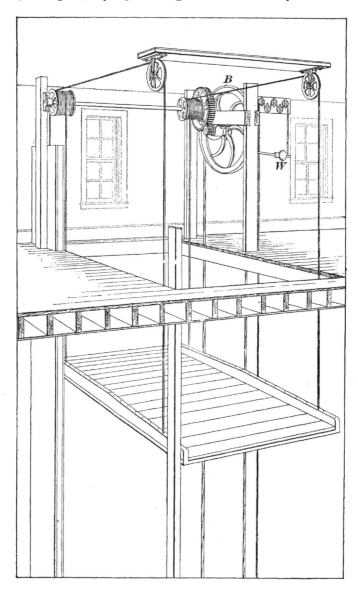

Fig. 19.

the brake is not applied automatically, a pawl P is thrown in mesh with the gearing when the elevator is at rest.

35. Motor Safeties. — Hand-power elevators have no *motor safeties*.

FIG. 20.

OPERATION AND MAINTENANCE.

36. The mechanisms of hand-power elevators are so simple that any one can operate them without difficulty; nevertheless, no less care should be exercised in handling them than power-driven elevators. Carelessness and neglect will prove just as dangerous with hand-power elevators as with any other type.

37. Hand-power elevators of the designs here illustrated cannot be operated from the car; a person riding on the same has no control over its movements and takes considerable risk. In operating, the operator lifts the brake and pulls the hand rope. In the design shown in Fig. 17, he must hold on to the brake rope, or **check-line,** as it is called, as long as the car is to move. This necessity is avoided by the arrangement shown in Figs. 18 and 19, the check-line passing over a number of small friction pulleys that give enough friction to the rope to hold the brake on or off after the operator has moved it by either an upward or downward movement of the hand. This device is a peculiarity of the hand-power elevators built by Morse, Williams & Co.

38. As the first and foremost rule it must be remembered that an elevator is built for a certain maximum load and that this load should never be exceeded.

39. All elevators should be started and stopped gradually. It takes more power to run an elevator up to a required speed than to maintain that speed thereafter; this additional starting power is the greater the shorter the time within which the necessary speed is attained, and the greater is also the stress in all parts of the machinery. Likewise, it takes considerable power to stop a moving elevator, which power is supplied by the braking device, and which is required to be the greater the quicker the elevator is stopped. Thus, if an elevator is stopped too quickly, enough stress may be put on the braking device to destroy it, causing accident. In hand-power elevators there is hardly any danger from quick starting, but there is from a sudden application of the brake, especially if the elevator is underbalanced and, as is often done, allowed to attain considerable speed in descending.

40. The drums, sheaves, and gears should be frequently inspected as to their fastenings to the shaft.

41. The brake needs particular attention, as the safety of the elevator depends on it.

42. If any car safety is provided, as there should be, the same should be examined frequently and carefully until the operator is satisfied that it is in good working condition; its parts should be kept well oiled and *should be kept clean*, to avoid their sticking and refusing to act in case of an emergency.

CARE OF WIRE ROPES, CABLES, AND GUIDES.

43. All that is said in Arts. **44** to **51** applies to all elevators, inasmuch as in all of them wire rope is used more or less and all have guides.

44. The wire ropes used in elevator work are made with hemp centers, to make them more pliable and thus more durable, on account of the short bends over comparatively small pulleys, or sheaves. Galvanized rope should not be used; the thin coating of zinc soon wears off, leaving the wires exposed to rapid deterioration by corrosion.

45. The wire ropes should be examined often and carefully; hoisting cables should be condemned as dangerous when the *wires* (not the strands) commence cracking. Wire ropes used for hoisting should under no circumstances be spliced.

46. Wire rope must be handled much more carefully than hemp rope, inasmuch as it is liable to kink and twist, which must be avoided on account of the harmful effect. Wire rope is best mounted on a reel that can be placed on a spindle to pay out the rope. If received from the supply house without a reel, the rope should be paid out by rolling the coil over the ground like a wheel. Wire rope should be lubricated like other moving machinery parts to preserve it,

To prevent rusting, raw linseed oil should be used and applied with a piece of sheepskin. The J. A. Roebling Sons Co. recommend to mix the linseed oil with the equal parts of Spanish brown or lampblack. The Otis Elevator Company recommend a mixture of 7 parts of linseed oil and 3 parts of tar oil. Another good preserving lubricant is made by heating and mixing well cylinder oil, graphite, tallow, and vegetable tar. When the **ropes,** or **cables,** as they are called frequently in elevator work, have once become well soaked, they need oiling only about every third or fourth month. They should not be allowed to become dirty and gummy.

47. In replacing worn ropes, particular attention must be given to the fastenings. In all cases where the ropes are replaced for the first time, it is best to carefully reproduce the joint as it was originally made by the makers or installators. An engineer taking charge of an elevator plant will, however, sometimes find rope fastenings of an inferior kind made by his predecessor. It may, therefore, be in order to call attention to the principal methods used by manufacturers.

48. The **shackle** used by the Otis Elevator Company is shown in Fig. 21. It consists of a split rod, the two legs A, A of which are bulged out and provided with noses at the ends. A collar B straddles the legs and eventually abuts against the noses. The rope is brought through the collar, bent over a **thimble** C, and passed back again through the collar, after which the free end is fastened by wrapping with wire. The wrapped end of the rope should be at least 8 inches long. The inside surfaces of the legs A and the outside surface of the thimble are concave to conform with the rope. Instead of the wire wrapping, clamps are sometimes used; the wrapping is to be preferred, however.

FIG. 21.

49. Another fastening is shown in Fig. 22. The rope is passed through a socket A forming part of the shackle; then it is untwisted for a short distance and the individual wires bent double. The socket is then filled with molten lead, or, better, with Babbitt metal, which should be of the best quality. The sockets should be warmed before the metal is poured to prevent chilling.

50. In fastening the rope to the drum, it must be observed that at the lowest position of the car the rope must still encircle the drum several times to reduce the stress at the point of fastening.

51. The guides should not be allowed to become gummy and should, therefore, be cleaned from time to time — about twice a month — and freshly lubricated.

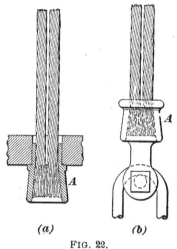

FIG. 22.

Gummy guides cause the car to alternately stick and free itself, making its motion jerky; and a bad case of sticking may cause the car to drop a distance great enough to break the cable and thus cause serious accident. In cleaning guides, a judicious occasional use of kerosene oil is recommended. For a lubricant on steel guides good cylinder oil is used; some use a composition that is seven-eighths cylinder oil and one-eighth plumbago, well mixed. Wooden guides are greased with No. 3 Albany grease or lard oil; a mixture of tar oil and wax is also recommended by some.

BELT ELEVATORS.

DEFINITION.

52. The term **belt elevator** is applied to that class of elevators that are driven directly by belts from line shafting, which shafting, in turn, may be driven by any prime mover and may be used for driving other machinery at the same time. Belt elevators are used for freight service principally, seldom for passenger service.

INSTALLATION.

53. The shaft from which the power is taken revolves

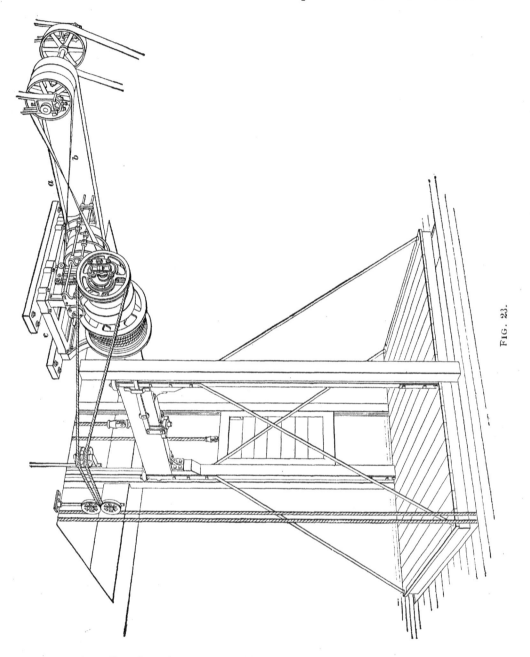

Fig. 23.

continually in the same direction independently of the motion of the elevator — that is, uncontrolled by the operator. The power is transmitted from this shaft to the elevator machine either direct, if the shaft is conveniently located, or by a countershaft. In either case, the shaft or countershaft carries a wide pulley that drives two belts, an open one *a*, Fig. 23, and a crossed one *b*. The elevator machine *c* is preferably placed on the ceiling, as shown, to save floor space, but it may be put on the floor as well. In many cases it will be possible to place it directly over the hoistway and thus save the expense of overhead sheaves.

GENERAL DESCRIPTION OF PARTS.

54. Following up the various parts again in the order named in Art. **2,** they present themselves as follows:

The *motor* of a belt elevator is simply a shaft carrying two loose pulleys and one tight pulley; it is designated by M in the various illustrations following hereafter.

The *transmitting devices* consist of either worm-gearing or spur gearing connecting the shaft M with the drum, worm-gearing being by far the more common arrangement.

When *counterbalancing*, worm-geared belt elevators are generally overbalanced; spur-geared ones are not. The reason for this is that worm-gearing works much smoother than spur gearing; it starts and stops gradually, offering much more resistance during the period of getting up speed, and acts as a kind of brake by itself in bringing the elevator to rest. The addition to the moving masses due to overbalancing, therefore, greatly increases the jerkiness of motion in spur-geared machines, while it has little influence that way in worm-geared ones.

The *controlling devices* consist of a pair of belt shifters, constituting the power control, and a brake, both being operated simultaneously by a shipper rope.

CONSTRUCTION OF CONTROLLING DEVICES.

55. In a belt elevator the controlling devices must be constructed in such a manner that the following requirements are fulfilled:

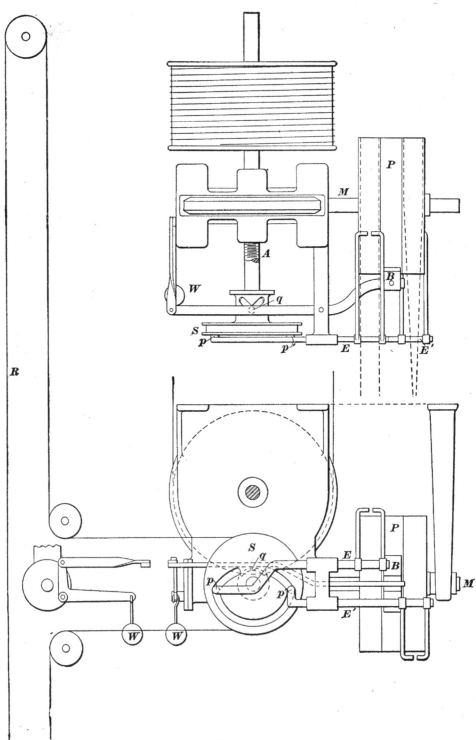

Fig. 24.

(a) When the shipper sheave is in the central position, both belts must be on their respective loose pulleys and the brake must be *on*.

(b) When the shipper sheave occupies the extreme right- or left-hand position, one belt must be on the tight pulley, while the other must be on its loose pulley, and the brake must be *off*.

(c) In their respective driving or non-driving positions, the belt shifters must be locked in place, so that they cannot be accidentally shifted.

(d) It must not be possible to throw the shipper sheave over too far.

(e) The central position of the shipper sheave must be distinctly defined, so as to give the operator warning against overthrowing the sheave from the one extreme position to the other. It is evident that this danger always exists.

56. The requirements stated in Art. **55** are met in various ways in practice. Fig. 24 shows in a diagrammatical way a typical arrangement for this purpose. The figure does not represent an actual machine, but was prepared to show the various elements of a belt elevator, separately explaining their functions. The shipper sheave S has two cam grooves into which enter corresponding pins p, p' on the ends of the belt shifters E, E'.

The cam groove, it will be noticed, has one concentric middle portion and two eccentric side portions.

When the sheave is in a central position, as shown, both belts are on their respective loose pulleys. A pull downwards on the shipper rope R swings the shipper sheave around to the left; the pin p of the left-hand belt shifter enters the left-hand eccentric portion of the cam groove and is thus forced to the right, while the other pin p' travels in the concentric portion of the groove and thus remains stationary. The open belt is thus shifted on to the tight pulley P while the crossed belt remains on its loose pulley; the elevator car then ascends. On pulling upwards on the rope R, the reverse takes place and the elevator car descends.

On the hub of the shipper sheave a **V**-shaped cam groove is formed, into which enters a pin g in the middle of a lever that carries on the one end a brake shoe B and on the other a weight W, which latter is so connected to it by means of a system of links that it tends to keep the brake on the tight pulley; a swing of the sheave either to the right or left lifts off the brake. Thus requirements (a) and (b), Art. **55,** are fulfilled. Requirement (c) is met by the shape of the cam groove, and not only are the shifters locked to the sheave in whatever position the same may be in, but also while one shifter is being moved the other is held positively and immovably in place by virtue of the concentric position of the cam groove.

Requirement (d) is met by the proper length of the groove, and requirement (e) by the sharp corner of the **V** groove, which will make itself distinctly felt to the touch of the operator.

57. As already said, the simple shipper rope is used for an *operating device*, special devices, such as described in Arts. **18** to **25,** being uncalled for, owing to the comparatively slow speed of belt elevators and to the fact that the car begins to move only after the belt has been shipped a considerable distance, so that it requires but little skill to complete the shift during the accelerating period of the car.

MOTOR SAFETIES.

58. Limit Stops on Shipper Rope.—In all elevators that are run by motive power the danger exists, if no provision be made against it, that, through the operator failing to arrest the car on time, the car or counterweight may be hoisted against the overhead work, causing damage and accident. Such danger does not exist in hand-power elevators with their slow speed, the resistance immediately being felt by the hand when the car strikes an obstruction. It is, therefore, one of the provisos in every power-elevator design that the power be shut off and the car be automatically arrested at the limit of its travel up or down.

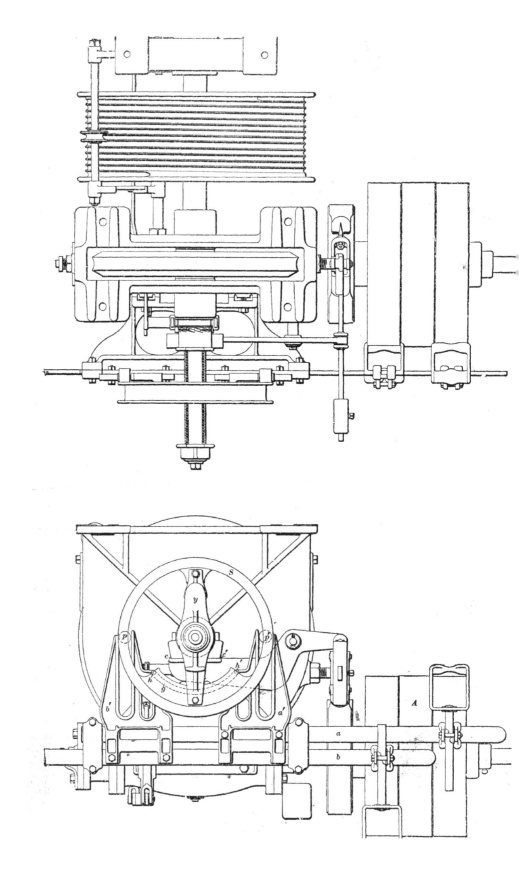

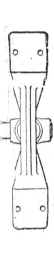
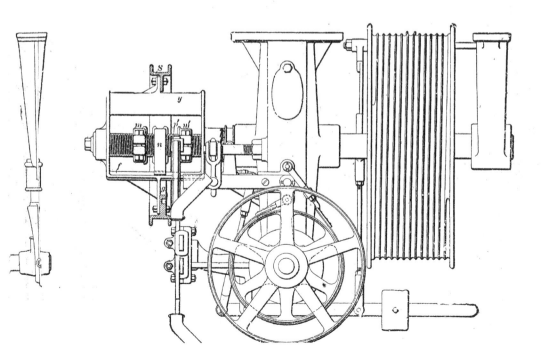

The means adopted for this purpose are called **limit stops,** and are of various designs. In all cases where a shipper rope passes straight through the car, knobs or buttons are clamped on the same, against which the car strikes when nearing its upper or lower limit of travel, thus operating the shipper sheave automatically. This means, of course, operates only as long as the shipper rope is intact. As it may easily occur that the shipper-rope connection is broken or the rope is otherwise ineffective, limit stops are also always provided on the motor itself.

59. Limit Stops on Motors.—For drum elevators the most common arrangement is that shown in Fig. 25. Let A

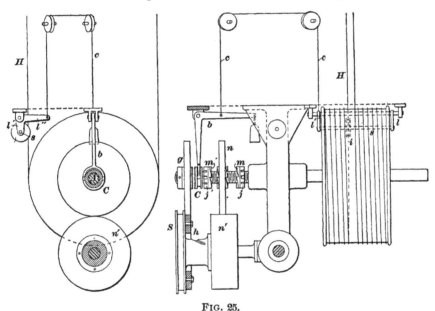

FIG. 25.

be a continuation of the drum shaft shown broken off in Fig 24. A screw thread is cut on this shaft and a gear-wheel n, the hub of which forms a nut, is placed on the threaded portion of the shaft; this gear-wheel meshes with another similar wheel n' bolted to the shipper sheave. It is evident that when the sheave is stationary and the drum rotates, the gear-wheel n will be prevented from revolving with the drum shaft, but will travel on the same in an axial direction either towards or from the drum, according to the sense of rotation of the latter. The hub of the wheel n has claws on either side, as shown, corresponding to similar claws formed on two other nuts m and m' that are clamped by jam nuts j and j', or in some other manner, securely to the drum shaft A. Now, it will be easily understood that when the wheel n travels either way, it will eventually be engaged by either one of the revolving nuts m or m' and be swung around, carrying with it the shipper sheave, with the effect of cutting off the power and applying the brake. The nuts m and m' can easily be adjusted to any position on the threaded portion of the drum shaft, and can thus be made to act when the car reaches the upper or lower limit in the hoistway.

SLACK-CABLE SAFETY.

60. Should the elevator car be obstructed in its descent by gummy guides or for any other reason and the motor continue to pay out the cable, the car would, if released suddenly, drop and most likely break the cable, causing damage; or should the car not drop, but be resting, for instance, on the bottom of the hoistway, the slack cable might still cause damage by getting into revolving parts of the machine. In any case, if the hoisting cable becomes slack, it will quickly be riding over itself on the drum or otherwise get entangled and must be straightened out again, which entails much labor and annoyance. A frequent occurrence is a slack cable produced by a careless handling of the shipper rope. It can often be noticed that when an operator in going up has missed his landing, he hastily reverses the machine to make his error good; the result is that the hoisting cable becomes slack. Now, most car safeties are so arranged that they will bind the car to the guides on the cables becoming slack. In his perplexity at the sudden stoppage of the car, the operator is likely to forget to shift the shipper rope so as to stop the machine, and the latter goes on paying out rope. Provision is made against such an emergency by contrivances called **slack-cable safeties.**

Fig. 25 shows the principle underlying such an arrangement: An idler i travels axially on its shaft s with the hoisting rope along the drum. The shaft s is supported on levers l, l' pivoted in a convenient manner. A cord c leads from the arm l'' of the lever l' over sheaves to a bell-crank b, one arm of which is weighted, while the other engages a clutch C. As long as the hoisting rope H is taut, the idler i is pushed outwards against the weight on the bell-crank b; but should the hoisting rope become slack, the weight on the bell-crank b will cause the clutch C to engage with a gear-wheel g mounted loosely on the drum shaft, and will cause the same to revolve with the drum shaft. The gear-wheel g meshes with another gear-wheel h fastened to the shipper sheave, so that the latter will be swung around when the hoisting cable becomes slack.

61. The principles illustrated by Figs. 24 and 25 are found embodied in all belt elevators in various ways.

EXAMPLES OF BELT ELEVATORS.

62. Fig. 26 is a plan, elevation, and side view of a worm-geared belt-elevator machine built by The Whittier Machine Company, Boston, Massachusetts, and designed to be placed on the floor. While there is no particular difficulty about understanding the operation of the machine, a few explanations will nevertheless be in order. The machine has two worms, one left-handed and one right-handed, actuating two worm-wheels that mesh together. This combination prevents the end thrust, which is unavoidable in single-worm machines and saves the power necessary to overcome the frictional resistance due to it. There is, consequently, also no wear to the end of the shaft, and no step bearing is required, as in single-worm machines.

With regard to the controlling devices, it will be noticed that the belt-shifter cam groove is continuous. Special provision is, therefore, made against throwing the sheave over too far by fastening a stop-plate T to the frame, and

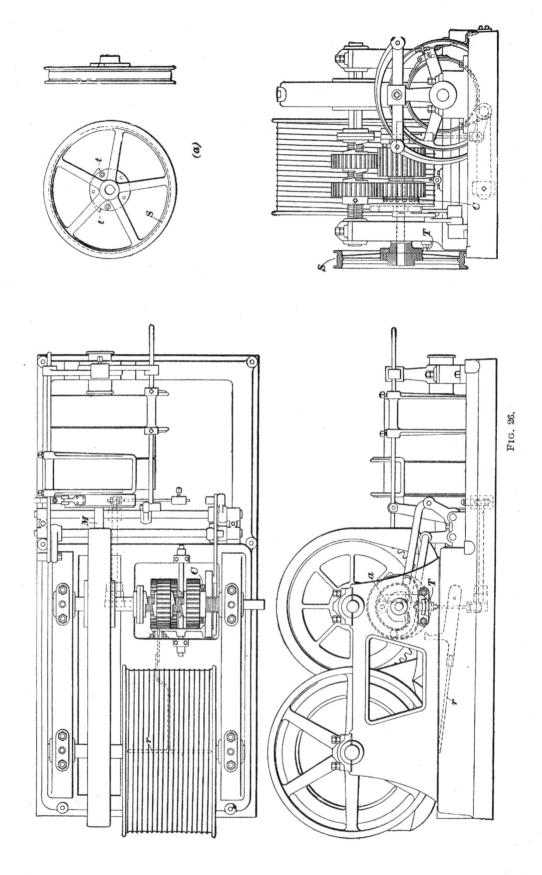

Fig. 26.

by stops t, t formed on the hub of the shipper sheave, as shown in detail in Fig. 26 (*a*).

The central non-driving position of the controlling device is made perceptible to the hand of the operator by nicking the brake cam, as shown in dotted lines at a in Fig. 26.

The limit stops are arranged practically in the same manner as in the typical drawing given in Fig. 25. The slack-cable safety is, however, radically different, inasmuch as the tension of the hoisting rope is not made use of; but, instead, the weight of one or more turns of rope hanging from the drum underneath in case the cable should become slack. For this purpose a rod r is placed across and underneath the drum, which rod is attached to the end of a weight-actuated lever that is tripped and closes the clutch C when there is any weight resting on the rod r. Both kinds of slack-cable safeties are extensively used.

63. Fig. 27 shows a worm-geared elevator built by Morse, Williams & Co., Philadelphia, Pennsylvania. In this machine the various requirements are fulfilled by slightly different means than have so far been shown, although the principles are the same. The difference lies in the manner in which the belt shifters are moved. Instead of the shipper sheave carrying the cam, the belt-shifter bars a and b have slotted cam pieces a' and b' attached to them, and the sheave carries two buttons, or projections, p and p'. It is easily seen that the effect is the same as when the shipper sheave carries the cam, with one advantage. As will be shown presently, it saves a good deal of complication in the way of gearing to put the shipper sheave loosely on the drum shaft instead of placing it on a separate stud, or shaft, in line with the shifter bars, as was done in the machines shown in Figs. 24 and 26.

As this, however, throws the center of the shipper sheave out of line with the shifter bars, the distance between must be bridged over. When using a cam on the sheave, this is done ordinarily by interposing double-arm levers l and l', as shown in Fig. 28, so that one advantage is gained at the sacrifice of another in that case. By arranging the parts as in Fig. 27, the necessity of the double-armed levers is avoided, making the machine so much simpler. Both types are in extensive use, however.

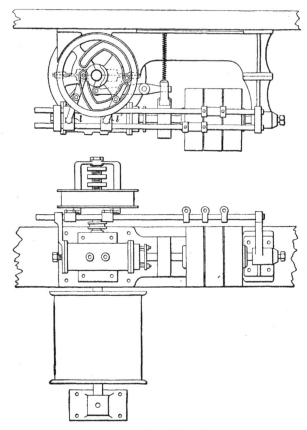

FIG. 28.

Reverting to Fig. 27, it will be seen that in turning the sheave to the right, for instance, the right-hand belt will be shifted on to the tight pulley A by virtue of the button p' entering the cam groove on the corresponding shifter-bar cam piece a'. The left-hand button p, however, will leave its cam b' entirely, and if no provision were made against it, the left-hand shifter would be unprotected, that is, liable to be shifted accidentally. To avoid this, that is, to lock the stationary shifter bar in place while the other is being moved (see requirement (*c*), Art. **55**), a circular groove g

is formed in the shipper sheave S, which fits over pins h and h' inserted in the shifter cam pieces a' and b'.

The central, or non-driving, position, as well as the extreme right and left positions, are strongly defined to the operator by the shape of the brake cam, which has a wide, flat surface for the neutral position and two smaller flat surfaces for the end positions, with the effect that when either of the corners c or c' of this cam passes through a position vertically below the center of rotation, the sheave will come to a quick and sudden stop.

64. It was mentioned in Art. **63** that by placing the shipper sheave on the drum shaft a simpler arrangement for the limit stops can be had. The usual arrangement is clearly shown in Figs. 27 and 28. The shipper sheave is provided with a yoke y, which takes the place of the hub. The yoke has formed on it a feather or rib f, upon which slides the traveling nut n that eventually engages with the fixed nuts m and m' in the manner already described. It is thus seen that the gearing shown in Figs. 24 and 26 is dispensed with. The yoke arrangement is in most extensive use and is found on almost every drum-elevator machine.

OPERATION AND MAINTENANCE OF BELT ELEVATORS.

65. The operation of belt elevators requires but little skill, the speed being comparatively slow; nevertheless, a certain amount of practice is required to "make the landings" exactly.

66. The operator *should never rely on the limit stops* to make a top or bottom landing, but should always operate the rope as at any intermediate floor. The limit stops are provided for an emergency and not for general use. They should be tried, however, once or twice every day, to see if they are operative and correctly set. The operator is to be cautioned against sudden reversals of the controlling device.

67. The brake needs adjusting from time to time. The necessity for this manifests itself by the car "settling" at the landings. A good deal of judgment is necessary in adjusting the brake; it should not grip too soon nor too late.

68. The belts should be kept under inspection and not allowed to become too slack. As new belts stretch considerably for a long period of time, they need closer attention than old ones.

69. Belt-elevator machines should be so installed, if possible, that the belts are not subjected to the influence of water, steam, or other moisture. In many cases, as in some factories, breweries, etc., this cannot always be done; it is then advisable to dress leather belts with a leather waterproofing compound, various brands of which are in the market, or to use rubber belts. Only first-class material should be used for elevator belts, especially if they are to be run in moist or damp places. Be careful to prevent oil from dripping on either leather or rubber belts, as the life of the same is greatly impaired thereby. Leather belting does not remain safe for any length of time in a temperature above 110° F.

70. In general, it is to be said that the elevator machine needs as much care as any other machine on the premises. It is too often considered of secondary importance and is neglected by the engineer in charge, especially as it is placed on the ceiling and more out of reach than other machinery. The bearings should be kept well oiled and the gearing should be kept clean. With regard to worm-gearing in particular, it may be well to mention that in the better class of machines it is enclosed in a casing, and the worm runs constantly in oil. This oil bath should be kept full and occasionally renewed to remove dirt and grit that may have accumulated. With new elevators this should be done more frequently than with the old ones that have been "run in"; with these fresh oil should be put in every two or three months.

71. Worm-gearing when new should, if possible, be less heavily loaded than when run in. A judicious observance of this rule is sure to prolong the life of the gearing considerably. Although conscientious manufacturers run in their worm-gearing before shipment, they can naturally do so only to a limited extent. It is said that the best oil to use on the worm bath is castor oil. The fact, however, that castor oil thickens when it becomes heated and that more or less heat is developed on worm-gearing, makes it desirable to use a mixture of 2 parts of castor oil and 1 part of the very best cylinder oil. Upon getting warm the cylinder oil runs freely, thus compensating for the property of castor oil mentioned.

72. Particular attention is to be paid to the lubrication of the thrust, or step, bearings of the worm, which should be renewed as soon as they show signs of cutting, since they will rapidly go from bad to worse. The step is generally made adjustable. The adjustment should be such that there is a little end play for the worm-shaft, say a scant thirty-second of an inch. This end play gives the oil a chance to enter between the bearing surfaces at every reversal of the worm.

If the steps are screwed up too tight, they will run hot at once and soon seize. The same as the worm and wheel, the step bearing requires to be run until a full uniform bearing surface is attained by a mutual adjustment through wear of the journal and its step. This mutual adjustment can be greatly facilitated by placing a leather washer behind the step.

73. Overhead sheave boxes must not be neglected. They should be kept lubricated with heavy grease in summer, with an addition of cylinder oil in winter.

74. Belt elevators should ordinarily not be run at a greater car speed than 60 feet per minute. The pulley speed should not exceed 400 revolutions per minute.

STEAM ELEVATORS.

CONSTRUCTION.

75. Motors.—Owing to the necessity of prompt starting, stopping, and reversing, the engines used for steam elevators are, without exception, duplex engines, generally of the vertical type. The cylinders are placed either on top or bottom, according to the kind of gearing used. Ordinary slide valves are used by some manufacturers, piston valves by others.

76. Transmitting Devices.—Practically all steam elevators are drum elevators. According to the kind of gearing used to connect the engine and drum, we can divide the elevator machines in general use into the following classes:

$$\text{Direct-geared elevators with} \begin{cases} \text{spur gearing.} \\ \text{worm-gearing.} \end{cases}$$
$$\text{Belt-geared elevators with} \begin{cases} \text{spur gearing.} \\ \text{worm-gearing.} \end{cases}$$

An example of an elevator of each class is given in Figs. 29, 31, 32, 33, and 34. The illustrations given do not by any means represent all the various designs of steam elevators, but have been chosen simply to illustrate the four types in most general use.

77. Counterbalancing.—Steam elevators are usually overbalanced when made with worm-gearing, but are not overbalanced when made with spur gears, for the same reason as belt elevators.

The engine thus furnishes full power only on the up trip of the car, and as the unbalanced portion of the car weight is generally considerable in steam elevators for the sake of safety, the engine must be of comparatively large capacity. To do away with the waste of energy in raising the unbalanced dead load of the car, and at the same time to avoid the use of the large moving masses of counterweights, recourse has been had occasionally to the expedient of connecting two cars with one hoisting drum, the ropes being so attached to it that when one car ascends the other descends, the motor furnishing power when the load on the ascending car is equal to or greater than that on the descending one. It is evident that the pressure on the bearings of the hoisting drum is equal to the weight of both cars plus their loads, and the stress in the hoisting ropes is equal to the weight of one car with its load. By connecting the two cars by a separate rope running over overhead sheaves, the bearings of the drum and the hoisting ropes are relieved of the car weights.

The obvious advantages of this double-car method of counterbalancing are, of course, gained at the disadvantage that both cars must move simultaneously.

78. Controlling Devices.—The controlling devices of a steam elevator consist of a steam-reversing valve and brake operated by a shipper rope, which is either a simple shipper rope or is in connection with some lever or wheel-operating device.

79. The steam valves used in the machine represented in Fig. 29, which is an **Otis spur-geared steam elevator,** are piston slide valves operated, as in most slide-valve engines, by eccentrics on the main shaft.

80. The reversing valve, which is shown in three positions in Fig. 30, reverses the motion of the engine by changing the piston valve of each engine from a direct to an indirect valve, and vice versa. Referring to Fig. 30 (a), the reversing valve a is shown in the position it occupies when hoisting. The valve is surrounded by live steam, which is also admitted to the cavity b. The live steam now flows through the ports c, c' into the passages e and e' leading to the ends of the piston valve, which now operates as a direct valve, taking steam at the ends and exhausting at the center. The exhaust from the piston valve passes through the passage d and ports o, o' into the exhaust passage f.

81. For lowering, the reversing valve occupies the position shown in Fig. 30 (b). Live steam now passes through

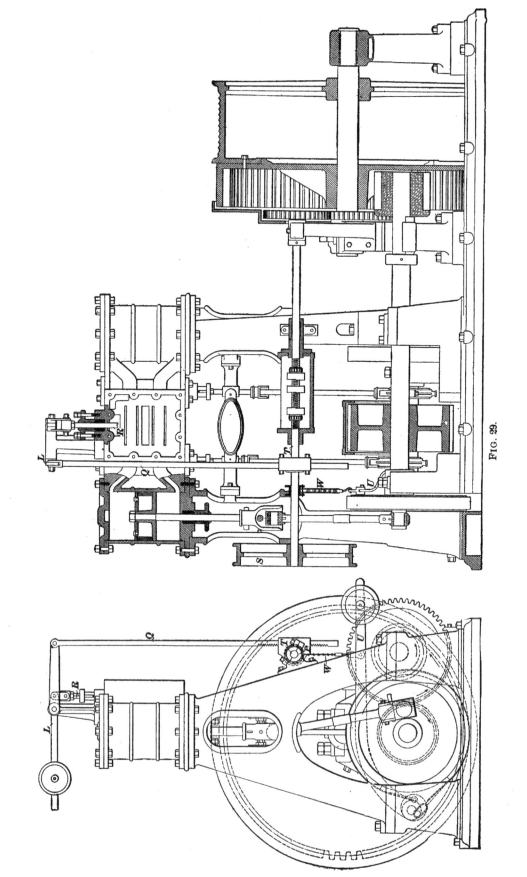

Fig. 29.

§ 37 ELEVATORS. 45

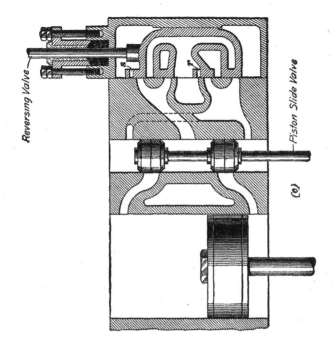
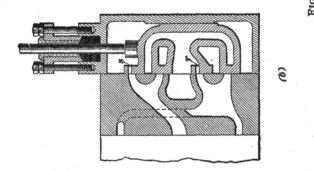
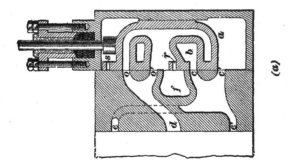

Fig. 30.

the small port s into the port o and thence through the passage d to the center of the piston valve, converting it into an indirect valve and thus reversing the motion of the engine. The exhaust from the engine passes through e and e' into the port c, and thence through the small port r in the valve to the exhaust passage f. The elevator being under-

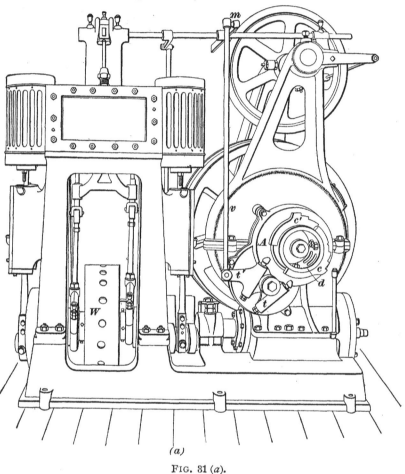

Fig. 31 (a).

balanced, the car descends by gravity, so that the steam ports need be uncovered entirely only for hoisting. Only enough steam to overcome the friction of the engine is needed in lowering, and for this reason the port s that admits steam for lowering is very small, being made up of a series of small holes drilled through the valve. The port r is constructed in the same manner.

82. When the car is at rest, the reversing valve occupies the position shown in Fig. 30 (*c*), where all ports are covered. A motion of the reversing valve in an upward direction will start the engine and hoist the car; a downward motion of the reversing valve will let the car descend.

83. The stem R of the reversing valve (see Fig. 29) is attached to a lever L that is actuated by a rack Q and

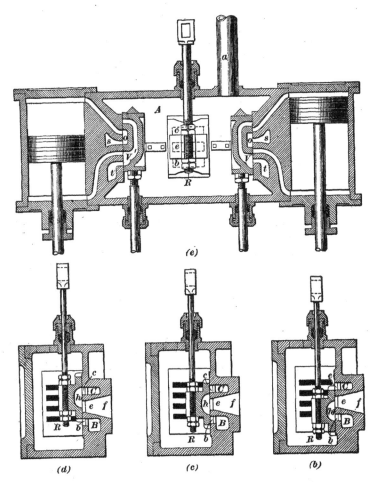

FIG. 31 (*b*), (*c*), (*d*), (*e*).

pinion P from the shipper-sheave shaft T, which latter is also connected to the brake lever U by a chain W. The operation of the brake is plain from the drawing.

84. The reversing valve in the machine shown in Fig. 31 (*a*), which is a **Crane worm-geared steam elevator,** is somewhat different from the Otis valve just described. Its action will be understood from Fig. 31 (*b*), (*c*), (*d*), and (*e*). Fig. 31 (*e*) is a vertical section through the two engine cylinders and the valve chest showing the steam-distributing slide valves V, V' in section and a front view of the reversing valve R, while Fig. 31 (*b*), (*c*), and (*d*) are transverse sections showing the three positions of the reversing valve. The action is as follows: Steam enters through the pipe a, Fig. 31 (*e*), the steam chest A. If the reversing valve R is moved to the position shown in Fig. 31 (*b*), the port c is opened, thus allowing steam to flow through the passage C into the cylinders, while the exhaust passes through passage B, port b, cavity h of the reversing valve, exhaust port e, and duct f into the atmosphere. To stop, the reversing valve is moved to the position shown in Fig. 31 (*c*), when it closes the passages B and C. To reverse the machine, the reversing valve is moved to the position shown in Fig. 31 (*d*). Steam then enters A through a, as before, but goes through port b and passage B to the distributing valves V, V', while the exhaust passes through passage C and port c.

85. The manner in which the engine is reversed may be explained as follows: The port c and passage C connect with the steam passage s, and the port b and passage B connect with the steam passage t in the valve seat of each engine. With the reversing valve in the position shown in Fig. 31 (*b*), the live steam is admitted through s into the central cavity o of the steam valves, while exhaust takes place through t. It is thus seen that the valves are now indirect. When the reversing valve takes the position shown in Fig. 31 (*d*), the live steam passes through b and B and through t into the steam valves, while now the exhaust steam passes through s into the cavity h of the reversing valve, and thence into e and f, and finally into the atmosphere. In this position the steam valves act as direct valves and give a motion to the engines contrary to that obtained when the valves act as indirect valves.

86. There is no brake shown on this machine. The worm having a sufficiently low pitch to make it self-locking, a brake is often dispensed with. When used, however, it consists of a wooden brake shoe, which is pressed against the wheel by means of springs and released by live steam.

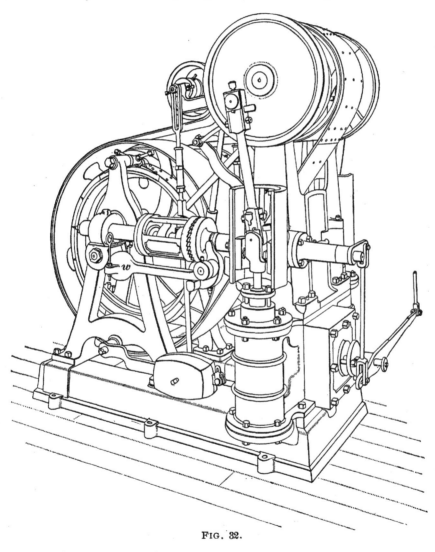

FIG. 32.

87. In the machine shown in Fig. 32, which is a belt and spur-gear steam elevator built by the Otis Elevator Company, a rotary reversing valve is used. Its action is much the same as that of the Otis reversing slide valve previously described.

88. Fig. 33 is an illustration of a belt and spur-gear steam elevator. The machine is built by the Otis Elevator Company of Chicago, formerly the Crane Elevator Company. The same kind of reversing valve is used in it as in

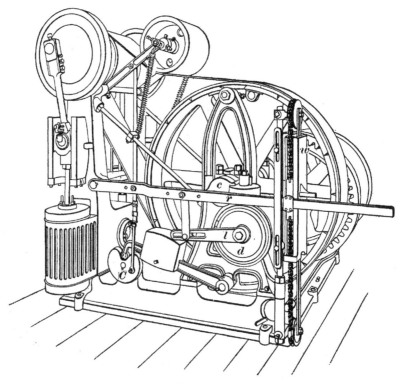

FIG. 33.

the machine shown in Fig. 31. Among the controlling devices in this machine is to be noted the heart-shaped brake cam C, the deep notch of which marks the central, or non-driving, position of the controlling mechanism.

89. Fig. 34 is an example of a belt and worm-geared elevator built by The Whittier Machine Company, now consolidated with the Otis Elevator Company.

90. Motor Safeties.—Motor safeties are provided in all cases, either in the shape of limit stops of the ordinary yoke type, as shown in Figs. 29 and 32, or of special design with the same underlying principle as shown in Fig. 31 (*a*).

91. The device used in the machine shown in Fig. 31 has some additional features, and is, therefore, shown in detail in Fig. 35.

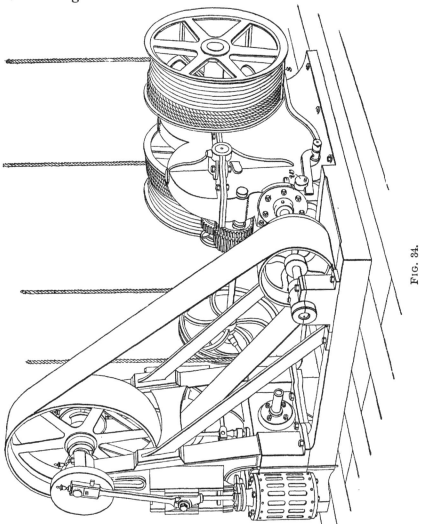

Fig. 34.

The winding-drum shaft carries a pinion p, Fig. 35 (*a*), meshing with a gear g. This latter gear has a second pinion p', which is solid with it and meshes with another gear g' mounted loosely on the winding-drum shaft. This gearing, which is similar to the back gearing of a lathe, is such that the gear g' will make less than one revolution for the whole number of revolutions of the drum shaft necessary to lift the car to the top. To the gear g' is attached a drum having long slots [see Fig. 31 (a)] into which are fitted adjustably the cams c, c', shown in Fig. 31 (a) and Fig. 35 (b), (c), and (d). These cams are located on the

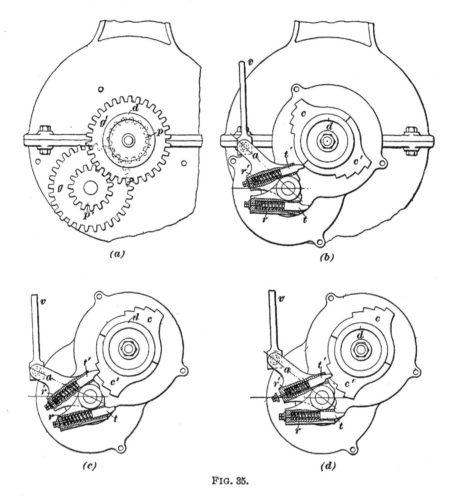

Fig. 35.

drum d in different planes and have two or more steps, as shown, and engage eventually each one of two spring-actuated triggers t, t' mounted in rockers r, r' on a stud and in planes corresponding to those of the cams. One of the rockers, which are rigidly connected to form one piece, has an arm a, to which is connected the rod v leading to the valve lever m, as shown in Fig. 31 (a). When the car reaches the top or bottom, respectively, of the hoistway, either cam c or c', as the case may be, engages its particular trigger t or t' and pulls or pushes the valve rod v.

Fig. 35 (b) shows the position of the various parts when the car is about midway in the elevator shaft. Suppose now the car to go up; then, as it nears the limit of its upward travel, the cam c' will engage the trigger t' with the first one of the steps and thus move the valve rod v until the trigger passes over the first step, as shown in Fig. 35 (c). This will slow down the car; on a farther motion of the car the second step of the cam will come into contact with the trigger, pulling the valve rod still farther, thus slowing down the car still more, and so forth, until the steam valve is entirely closed and the car stops.

92. The gradual choking off of the steam supply by the successive steps has the effect that with a heavy load on the car the latter will finally reach the top very much slower than with a light load, and may even stop short of the last landing. Conversely, if the apparatus is so adjusted that with a heavy load the car will finally stop at the lowest landing, it may do so with a light load, but only very slowly and even may not reach the landing. The apparatus illustrated in Figs. 31 and 35 provides for these conditions, inasmuch as it permits the operator on the car to operate the controlling device to some extent even after the automatic stop has performed its function. This is accomplished by making the triggers t, t' spring actuated. The springs will not yield to the action of the cams, the pressure being a transverse one, but they will yield if a pull or push, respectively, is exerted on the valve rod v by the operator. Thus, if it is found, for instance, that after the cam c' [see Fig. 35 (c)] has acted upon the trigger t' so as to slow down the car, the latter moves too slowly, the operator may pull on the rod v and bring it into the position shown in Fig. 35 (d), the spring of said trigger permitting this, and thus partly reopen the steam port.

93. The machine shown in Fig. 33 has a different arrangement for an automatic stop. Back of the disk d is a plate revolved from the drum shaft by a worm and gear (covered in the illustration by the case c). This plate has a spiral groove that moves a stud connected to the disk d in and out until it strikes and touches on adjustable stops and causes the disk d and lever l to turn with it, thus centering the reversing lever r.

94. Slack-Cable Safety.—Slack-cable safeties are generally provided on all steam elevators, although shown only in Figs. 32, 33, and 34. They consist in all cases of a rod running across the under side of the winding drum and so arranged that it is depressed by the weight of any loose cable that may form. When so depressed, it releases a spring or weight, which, in turn, acts upon the controlling device, shutting off the steam. The aforesaid rod is seen in Figs. 33 and 34 at s and the weight at w. In Fig. 32, the weight w, when released, throws in a clutch that causes the limit-stop yoke to turn the valve rock shaft.

OPERATION AND MAINTENANCE.

95. The operation of a steam elevator is exceedingly simple and at once familiar to every one able to run a steam engine. Too great care cannot be bestowed on the hoisting ropes and the various safety appliances; the weights handled by steam elevators being usually great, the risk to human life that is incurred by neglect on the part of the engineer is correspondingly great.

96. In belt-geared elevators, attention must be paid particularly to the driving belt. A breakage of this belt, it will be observed, throws the car and all there is upon it on the car safeties with disastrous results if the latter should prove inefficient. An elevator belt performs a duty much more severe than ordinary belting; it runs over a large and a small pulley and under an idler to give it as great as possible an arc of contact on the small pulley; it must also run in opposite directions alternately, so that there is always considerable slip. Such a belt should, therefore, be of the best quality obtainable and should be well cared for. The leather used should be genuine oak-tanned stock; the pieces should be cut from the hide in such a way that the hide center will be the center of the belt. The pieces should be well stretched before being made up. They should not be more than 50 inches in length, including laps, and should be joined by a so-called *lock lap*, making a perfect joint. A *straight lap should not be used under any circumstances.* Besides being of best quality, the cement used must be very pliable on account of the short turn of the belt under the idler and over the small pulley. The belt should be riveted as a precaution against a lap becoming loose, so that the rivets may hold the defective lap together until it is discovered and repaired.

Lacing belts must not be resorted to, as the laces soon break, due to running over the small pulley.

It is recommended by elevator men to give the belt an occasional dressing with castor oil to keep it pliable.

ELEVATORS.

(PART 2.)

ELECTRIC ELEVATORS.

INTRODUCTION.

1. Treating elevators in the order of their development, the hydraulic elevator would follow after the steam elevator, because the electric elevator is the latest competitor in the field. Nevertheless, as most electric elevators are of the drum type, and therefore similar in many ways to hand, belt, and steam elevators, they will be considered before the older type.

INDIRECT-CONNECTED ELECTRIC ELEVATORS.

BELT-CONNECTED, BELT-SHIFTING ELEVATORS.

2. The first mode of application of the electric motor to elevator machinery was simply a substitution of an electric motor for whatever kind of power was previously used for driving the line shafting of an ordinary belt elevator. The motor was started by an ordinary main switch and starting box and ran continuously in one direction, the elevator being controlled in the same manner as other belt elevators. If such an elevator is not in constant use, the electric motor must be stopped and started frequently, which, with an ordinary switch and hand starting box, compels the operator to go to the starting box every time the elevator is used. To avoid this, the switch and starting box are operated by a hand rope running through the car in the same manner as the shipper rope, or to avoid the handling of the two ropes, the shipper rope may serve both for shifting the belts and for operating the switch and rheostat.

BELT-CONNECTED, BELT-SHIFTING, REVERSIBLE-MOTOR ELECTRIC ELEVATOR.

3. General Description.—By introducing a reversing switch instead of the single switch, the motor can be reversed by reversing the current in the armature. The necessity for two belts, an open and a crossed one, is then obviated, and one belt between the countershaft and elevator machine is sufficient, this belt being shifted from a loose pulley to a tight one to start the car in either direction, that is, up or down.

4. Belt-shifting electric elevators being nothing but combinations of belt elevators with an electric motor, we can confine our remarks with respect to the various parts of these elevators to motors and controlling devices, all the other parts being the same as in ordinary belt elevators.

5. Motors.—For belt-shifting elevators, continuous-current, constant-potential, shunt-wound, single-speed motors are generally used, and since the motor starts without load, no rush of current that might injure the armature takes place at starting. Any kind of alternating-current motor may be used for belt-shifting elevators when the motor runs continuously. When, however, the motor is to be stopped and started frequently, polyphase synchronous motors or induction motors must be used, because these motors will start by themselves, while single-phase motors will not.

6. Controlling Devices.—Aside from the belt shifters in belt-shifting elevators, the power control consists of a switch and a rheostat. For combinations in which the motor runs continuously in one direction and is started and stopped only occasionally, the ordinary switch and starting box operated by hand are sufficient. If, however, the switch and rheostat are to be operated by a hand rope or other operating device from the car, special mechanisms become necessary, since the simple pull on the hand rope cannot give the necessary motions. To prevent a possible damaging rush of current in starting such an electric motor as is used in elevator work, the main switch is closed with all the starting resistance in the armature circuit, which resistance is then gradually cut out as the speed of the motor increases, until the motor is finally (when running at its normal speed) connected directly to the mains. After stopping, this resistance should all be in again, so as to make the apparatus ready for the next start; and since starting may follow quickly upon stopping, this restitution of the apparatus to its starting conditions after stopping must be effected quickly. When the switch and starting box are manipulated by hand, the above requirements can be easily fulfilled, but not when they are operated together from a hand rope. To obtain the required motions, various contrivances have been devised and are largely used. A few examples are given.

7. Mechanically Operated Rheostats. — The most natural way to

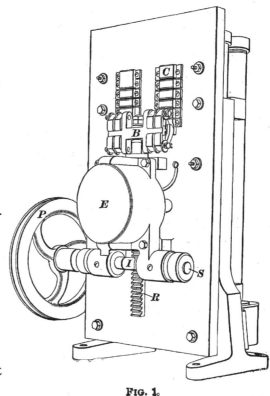

FIG. 1.

gradually cut out the starting resistance as the speed of the motor increases is to mechanically connect the starting box to the motor shaft. Fig. 1 shows an apparatus made by the Automatic Switch Company and designed to be used with motors running always in one direction, that is, in our case, with an indirect-connected or belt-shifting, non-reversible elevator machine. The pulley P is belted to a smaller pulley on the motor shaft or countershaft and drives a shaft S having formed on it a two-toothed pinion I. When the motor is running, a rack R is drawn into mesh with the pinion I by means of an electromagnet E energized by a coil in shunt with the motor circuit. As soon as the circuit is closed and the motor commences to revolve, the rack ascends and with it the contact bar B that is carried on its upper end. The contact bar passes successively over the contacts C, gradually cutting out resistance. As soon as the current is broken, the magnet is deenergized and the contact arm drops back, the rack R springing out of gear with the pinion.

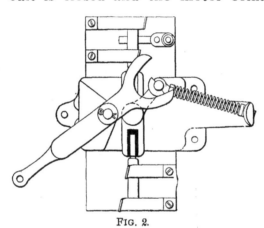

FIG. 2.

8. In connection with the starter shown in Fig. 1, a simple snap switch is used, such as is shown in Fig. 2; the action of this will be readily understood. It is operated either by hand or by a separate hand rope or cord running parallel to the shipper rope in the hoistway.

9. Fig. 3 is a diagram of an installation using the starting box shown in Fig. 1. Fig. 4 is a diagram of the connections; this will prove useful to engineers wishing to drive existing belt elevators by an electric motor.

10. In case a belt-shifting elevator is to be run with a single belt, the motor must be reversible. A **reversing**

switch is then used instead of the single snap switch shown in Fig. 2. Such a reversing switch, made by the Automatic Switch Company, is shown in Fig. 5, which also gives a diagram of the connections. The reversing switch has

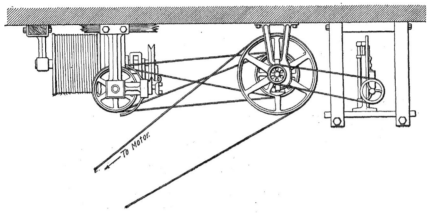

FIG. 3.

four sets of contacts a, b, a', b', each consisting of three clips, and two blades B and B', which are insulated from each other. The clips are connected with the terminals of the various parts (motor armature, field, resistance,

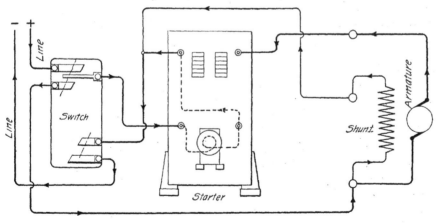

FIG. 4.

and starter magnet), as shown. When the switch is pulled up, blade B connects the three clips at a and blade B' connects the three clips at b. This allows the current to flow through the armature, the shunt field, and the resistance, and the elevator ascends. When the reversing switch is pulled down, B connects the three clips at a' together and B' connects the three clips at b'. This reverses the flow of the current through the armature, because the wires on the switch that connect the upper and lower horizontal clips are crossed; the current in the shunt field flows in the same direction, no matter whether the switch is up or down; hence, pulling down the switch

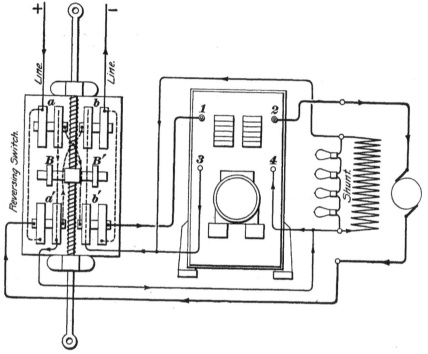

Fig. 5.

reverses the motor. The terminals of the armature resistance are shown at *1* and *2*; *3* and *4* are the terminals of the magnet that throws the rack into and out of gear. With this explanation the student will be able to trace the path of the current without difficulty.

11. It is often observed on opening the circuit that there is considerable sparking at the clips connected to the shunt field. This is due to the self-induction of the field. To reduce this sparking, it is a good plan to connect across the shunt a series of incandescent lamps having a combined voltage of from 6 to 8 times that of the line current; that is, in case a 110-volt lighting current is used, a series of, say, four 220-volt lamps is inserted, through which the induction current of the field is discharged. Since the starter is belted to the machine or countershaft, it will be reversed with it; it must, therefore, be so arranged that it will lift the crossbar B, Fig. 1, no matter in which direction the motor runs. This is done in this kind of starter by substituting for the two-toothed pinion I an eccentric operating a pawl. Otherwise the "reversible starter" is the same as the "non-reversible" one.

12. Another kind of mechanically operated starter is shown in plan and elevation in Fig. 6. It is made and patented by the Otis Elevator Company. Its action is different from the apparatus described in the foregoing article in so far that it is not connected mechanically to the motor or countershaft but to the main-switch spindle, and the gradual cutting out of resistance is obtained by a dashpot. The following description is taken from the patent specifications:

A box A contains in its rear part A' resistance coils, and in its front part the operating mechanism, the essential features of which consist of a snap switch 1, an arm 2 for operating the snap switch, and a brush-carrying arm 3, which brush operates in connection with a resistance device 10; the brush arm 3 is, in the present instance, provided with a counterbalance 9 and controlled by a dashpot 4; arm 3 is mounted on a shaft 5, by means of which it is operated in the manner described later.

The switch 1 comprises essentially a knife blade 7, mounted on a pivot 6, adapted to engage and disengage the contacts $8, 8$, and connected to this knife is a cam 16 having a notch 17, into which projects the end of the arm 2 for moving the cam; the cam is further provided with recesses and projections 18, with which a spring catch 15 cooperates, under the stress of a spring $15'$ for holding the switch

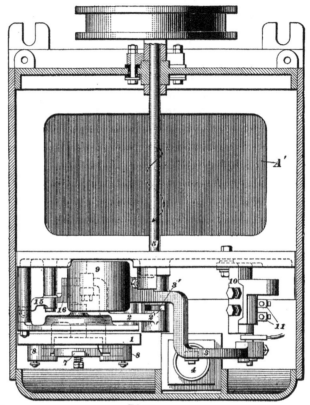

Plan

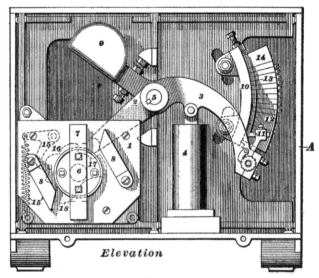

Elevation

FIG. 6.

in different positions and for making it complete its movements after it has been started, so as to produce the sudden engagement and disengagement in the manner well known in connection with snap switches. The arm *2* is rigidly connected to the shaft *5* so as to move therewith, while the brush-carrying arm *3* is loosely mounted on the shaft *5*; interposed between the two arms is a catch, or stop, so arranged that the arm *2* may move independently of the arm *3* when the parts are in one position, but when it is moved in the opposite direction and the arm *3* is in another position, they will move together. This catch consists of a projection *2'* on the hub of the arm *2* working in a slot *3'* in the hub of the brush-carrying arm *3*.

The brush-carrying arm *3* carries a brush *11* adapted to bear on the resistance-contact device *10*, and the contacts are arranged so that the contact *12* will permit a considerable movement of the brush before any of the resistance is cut out. While the contacts *13* are connected by the resistances in box compartment *A'* in the usual way, the contact *14* is connected directly with the line; so that while the brush is on the contact *12* all the resistance is included in the circuit, and as it sweeps over contacts *13* more or less of the resistance is cut out until it bears on the contact *14*, when all the resistance is out of the circuit. This resistance device *10* is made on the arc of the circle and is adjusted in the box by means of lugs and bolts engaging slots in the frame of the box.

In the figure, the circuit is shown open and all the resistance is included in the circuit, the catch *2'* bearing on one side of the slot *3'* of the brush-carrying arm *3*, holding the parts in the position shown. If, now, the shaft *5* is turned in the direction of the arrow, that is, to start the motor, the arm *2* operating through the cam *16* will move the switch blade *7* so as to engage the contacts *8*, the spring catch *15* riding over the projection *18* and tending to complete the throw of the switch arm as it enters the adjacent depression on the other side of the projection *18*, making a snap switch. The catch *2'* moves through the slot *3'* and leaves the brush-carrying arm *3* free to move, which, under the influence of the counterbalance *9*, it commences to do at once, but its movement is retarded more or less by the dashpot *4*. The parts are so arranged that before the brush *11* moves off the resistance contact *12*, the switch *1* has closed the circuit through the contacts *8* and the brush-carrying arm moves gradually over the resistance contacts, cutting them out, until the brush *11* bears on the contact *14*, by which time the motor has come up to speed. When the shaft *5* is turned in the direction opposite the arrow, that is, to stop the motor, the projection *2'* bears on the side of the slot *3'* so that as the arm *2* is turned to open the switch *1*, the brush *11* is moved over the resistance contacts, insuring the inclusion of the resistance in the circuit. It will be noted that the slot *17* in the cam *16* is of such dimensions as to permit the inclusion of a greater part of the resistance contacts before the knife blade *7* is actually moved from the contacts *8*.

13. Solenoid Rheostats.—Instead of the weight *9*, Fig. 6, a solenoid is used in many starting devices. This permits the rheostat to be mounted separate from the switch, no mechanical connection between the two being required. The switch alone is mechanically operated by the hand rope or other operating device. Fig. 7 shows one form of solenoid rheostat, as manufactured by the Elektron Manufacturing Company. The armature current enters at the binding post *1*, whence it goes to the contact arm A, through the series of resistances R, and out at the binding

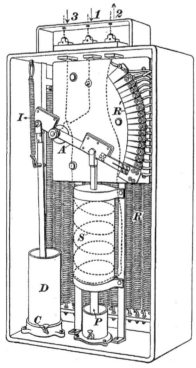

FIG. 7.

post *2*. The solenoid current, taken from the main switch, enters at binding post *3*, goes through the windings of the solenoid S, and leaves at binding post *2*. As soon as the main switch is closed, the solenoid is energized and draws in the iron plunger P, raising the arm A, and thus making the contact piece at the end slide over the sectors R' of the rheostat and cutting out resistance from the armature circuit. In order that this may be effected gradually, the other end of the arm A is connected by a rod with a piston fitting in a dashpot D. In moving downwards, this piston must displace the air in the dashpot, and the speed with which this may be done is regulated by the stop-cock C. To bring the apparatus back to its original position at the breaking of the circuit, the piston end of the arm is provided with a spring I that is put in tension while the resistance is cut out. On opening the circuit, the spring pulls up the arm and dashpot piston, and in order that this may be effected quickly the dashpot has a relief valve that will open while the piston is going up.

14. The apparatus described in the foregoing articles as applicable to belt-shifting elevators are used for a number of other purposes, among which their connection with electrically driven pumps for hydraulic elevators is of special interest.

DIRECT-CONNECTED ELECTRIC ELEVATORS.

DIRECT-CONNECTED, BELTED ELECTRIC ELEVATOR.

15. The second step taken in the development of the electric elevator was the elimination of the countershaft and the tight and loose pulley, and the substitution therefor of a belt connecting the motor directly with the elevator machine. The mechanisms used in belt elevators for shifting the belt then became superfluous. Although the elimination of the countershaft seems a small and natural step to take, it makes a great change in the working conditions of the elevator, since in the belt-shifting types the motor starts without load, which is applied only after the motor has attained its normal speed; while in the direct-connected type, the motor must start under load. There is nothing gained by having the motor and the elevator separate and belted together, and therefore direct-connected belted elevators are never used; they are described here only to help us to arrive gradually at the form of elevator now commonly used.

DIRECT-CONNECTED ELEVATORS.

16. Connection of Motors and Machines.—The working conditions of the direct-connected belted elevator are not changed when the motor is coupled directly to the shaft of the elevator machine, and in the modern type of electric elevator this is always done, the motor being mounted on the same base with the machine.

17. Motors.—Since in direct-connected electric elevators the motor must start under load and must, therefore, have a strong torque, it must also get up speed rapidly though gradually. Of these two conditions the last-named one is fulfilled by peculiar controlling devices that are described below, while the first-named one is fulfilled by the construction of the motor itself, which is generally of the compound-wound type—a series-field coil serving to give the necessary torque at starting and the shunt coil steadying the field. The series coils are generally cut out when the motor has attained normal speed, after which the motor runs as a simple shunt-wound motor.

Of alternating-current motors, only the two-phase or three-phase induction motors prove satisfactory for direct-connected electric elevators, since they will start under load with sufficient torque. These motors behave, as far as their action in the elevator combination goes, just like shunt-wound continuous-current motors.

18. Transmitting Devices.—The transmitting devices between the motor and car consist, with few exceptions, of worm-gearing, drum, and rope. The worm-shaft is almost invariably coupled to the motor shaft by a flange coupling, serving at the same time as a brake pulley. Both single worm- and double worm-gearing are used, as will be seen from the illustrations given farther on, the double worm being used mostly on heavy machines, to avoid the end thrust of the worm-shaft. Such heavy machines are also frequently provided with back gearing. Ordinarily, however, single worm-gearing is used, great care and ingenuity being displayed in the design of the step bearings for the worm.

19. Counterbalancing.—Direct-connected electric elevators of the drum type are always overbalanced.

20. Controlling Devices.—The power control of direct-connected electric elevators is entirely electrical, there being no belts to shift or similar mechanical operations to perform; but, besides breaking the current, the motor must be reversed. Hence, besides the simple snap switch and rheostat already mentioned in connection with belt-shifting electric elevators, a **reversing switch** or **pole changer** is needed.

In elevator practice, the complete apparatus necessary to control the electric motor—the **power control,** as we have called it—is called a **controller,** especially if the various parts of it are built together in such a way as to make a separate, self-contained piece of machinery. A number of different forms of such controllers are used by the various manufacturers of electric elevators, and they will be described with the various designs shown.

21. Brakes.—The braking arrangements used are either entirely mechanical, that is, such as are used in connection with belt and steam elevators, or electrical mechanical, or wholly electrical.

22. Operating Devices.—In the majority of electric elevators the operating devices are mechanical, such as hand ropes, hand wheels, and levers. Electrical operating devices

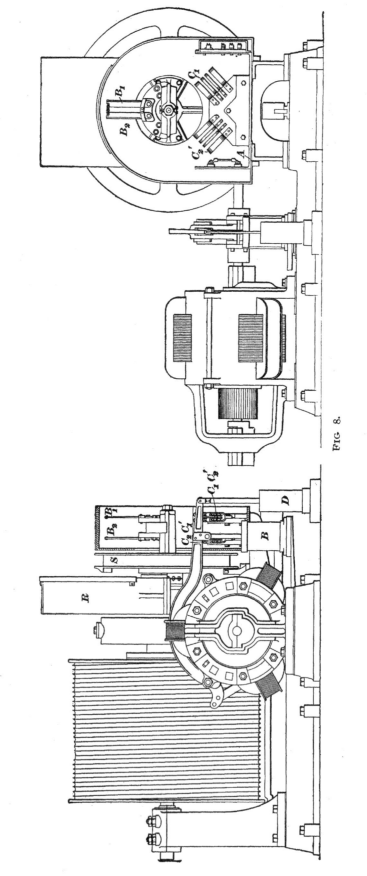

Fig. 8.

are being introduced, however, with success in connection with the magnet system of control, which is described later.

23. Motor Safeties.—Motor safeties are used in various forms; they are either mechanical or electrical or both.

EXAMPLES OF ELECTRIC ELEVATORS.

INTRODUCTION.

24. The examples of electric elevators here given do not represent all the various designs in the market, nor does the order in which they are described indicate any superiority of design of one make over another. A careful study of these will give a person enough insight into the construction and operation of this class of machinery to enable him to handle other makes of machines.

ELEKTRON ELEVATORS.

25. Motors.—Fig. 8 is an end and side elevation of an electric elevator made by the Elektron Manufacturing Company. The motor is the well-known Perret multipolar machine, shunt-wound.

26. Transmitting Devices.—The transmitting devices are single worm-gearing, drum, and rope. The arrangement of the step bearing of the worm is shown in Fig. 9. Alternate phosphor-bronze and steel disks are used to distribute the wear. The worm-shaft is attached to the motor shaft by means of a flange coupling F, which serves at the same time as a brake pulley.

27. Simple Controller.—The Elektron Manufacturing Company uses various kinds of controllers for various kinds of elevators. The simplest arrangement used is a double-throw switch attached to the hub of the shipper sheave S, Fig. 8, and a solenoid rheostat placed anywhere conveniently near the machine; such a rheostat is shown in Fig. 7 and another form in Fig. 10.

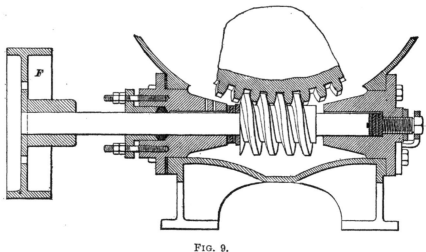

Fig. 9.

The switch consists of a casting A, Fig. 8, supported on the frame of the machine and carrying four sets of clips C_1, C_2, and C_1', C_2', to which the necessary line, field, armature, solenoid, and electric-brake connections are made as shown below. The switch blades B_1, B_2 attached to the shipper sheave engage the clips C_1, C_2, or C_1' C_2' for the up trip and the down trip, respectively. In Fig. 8 the blades are shown in their neutral position; that is, when the elevator is at rest. It will be seen that to start the elevator up or down, the sheave with the blades must be turned through an arc of 135°, the clips being set at right angles. This long travel is given for the purpose of giving the rheostat arm time to fall back into its starting position before the current in the armature can possibly be reversed; it also helps to reduce sparking and flashing at the clips.

28. Ordinary Brake.—The brake used in these machines is, for ordinary service, a simple mechanical one, which is released by a cam on the shipper sheave through a system of levers and applied by a weight, as with belt elevators. For passenger service, an electrical-mechanical brake is used, which is released by an electromagnet and applied by gravity. This arrangement is shown in Fig. 8, in which the brake magnet is marked B; the rapidity of action of the same is regulated by a dashpot D.

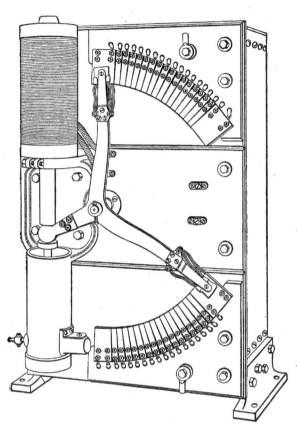

FIG. 10.

29. Fig. 11 is a diagram of the electrical connections between the switch, rheostat, brake, and motor. It will be useful to follow out these connections. The lines are connected through the fuses f, f and the double-pole switch s to the elevator switch at the binding posts L_6 and L_7. Supposing the blades of the switch to be thrown to the right, that is, across the clips C_1 and C_2, and the current to enter at the binding post L_6, then it passes first to clip 1 of the set C_1, whence it divides by means of the switch blade among the clips 2, 3, and 4. From 2 it passes to binding post L_5, thence through the field windings of the motor, back to the binding post L_4, thence to the clip b of set C_2,

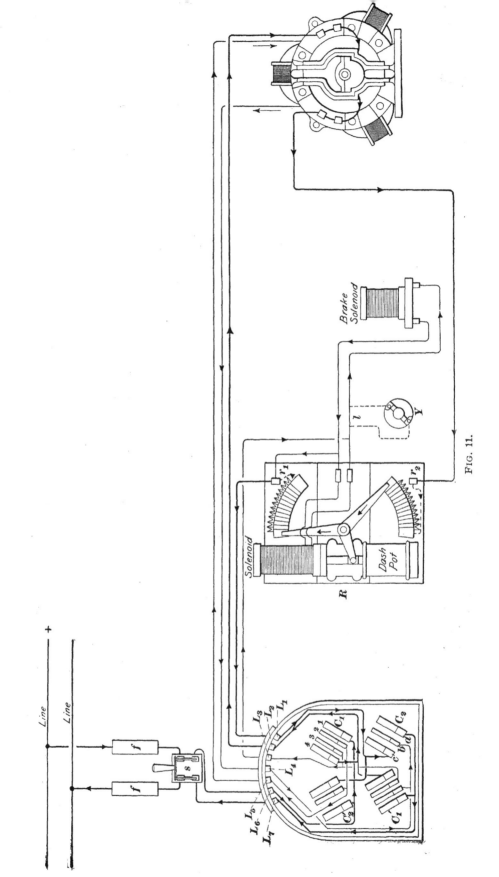

Fig. 11.

over the blade crossing this set of clips to clip a, thence to binding post L_7 and to the line, thus completing the shunt circuit for the field. From clip 3 the current goes to the binding post L_3, through the solenoid windings of the rheostat R to the binding post r_1 of the rheostat to the binding post L_1 of the switch, to clip c of set C_2 over the blade to the clip a, to the binding post L_7 to the line, thus completing the circuit through the solenoid. From clip 4 the current goes to binding post L_2, thence through the armature of the motor to the binding post r_2 of the rheostat, through the lower half of the resistance, through the rheostat arm and the upper half of the resistance to binding post r_1, to L_1, c, a, L_7, and line, thus completing the armature circuit. Throwing the blades to the left, we will find, in following out the three circuits again, that the current traverses the field circuit in the same direction as before, but that the current in the armature is reversed, thus reversing the motor. The electromagnet windings of the brake are in shunt with the solenoid circuit, as is easily seen from the diagram.

30. The operation of this elevator is as follows: When the shipper sheave is thrown over to the right or left, the brake magnet is energized and tends to slowly release the brake, since the dashpot prevents too sudden a release; at the same time the solenoid is energized. This tends to slowly cut out the resistance from the armature circuit; the dashpot prevents too quick an action, and it is so adjusted that all the resistance will be cut out by the time the motor reaches its normal speed. Upon breaking the circuits, the brake is at once applied and the resistance arm drops back into its original position, ready for another start.

31. Dynamic Brake.—On high-speed elevators, in order to get a particularly smooth stop, the Elektron Manufacturing Company uses, in addition to the electrical-mechanical brake, a so-called dynamic brake, which, indicated in Fig. 8 at R, is usually placed on a bracket between the shipper sheave and worm-gear case. It is shown in detail in Fig. 12 and consists of a switch lever L, actuated by a cam on the operating sheave, and a variable resistance.

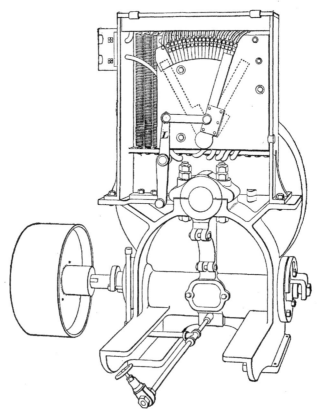

FIG. 12.

This resistance is so connected to the system that the armature is short-circuited through it immediately after the circuit from the line is broken to stop the elevator, thus acting as a stopping resistance, the motor acting as a dynamo and sending a current through the resistance. This has the effect of slowing the motor down quickly but smoothly, like a brake, and more smoothly than an ordinary frictional brake. The smoothness of the stop is made still more marked by the resistance being gradually cut out of the armature short circuit as the motor slows down, the cam operating the lever L being so constructed as to first cut in all the resistance at the instant the main circuit is broken; on being turned farther by the operator, the switch lever is caused to brush over the resistance contacts, thus gradually cutting the resistance down to zero. Of course this short circuit is opened before the elevator is started again. As has been said, the dynamic brake is used only in addition to the ordinary brake, the latter being necessary to hold the car stationary after it has been stopped.

32. Fig. 13 shows diagrammatically the connections when the dynamic brake is used. The field must necessarily remain excited after the armature circuit is broken and the armature short-circuited, in order to make the motor act as a dynamo. The field is, for the sake of simplicity, kept excited all the time, but in order to cut down the current thus constantly wasted while the elevator is standing still, a resistance is inserted in the fields. When the elevator is started, this resistance is short-circuited, thus giving the fields the full current due to its windings and, consequently, the full torque available. When the elevator is stopped, the resistance is cut in, choking the field current, but leaving it strong enough to give sufficient magnetism to get a dynamic-brake effect.

33. Speed Regulating Controller.—Another type of controller used by the Elektron Manufacturing Company is shown in Figs. 14 and 15, while the diagram of connections is given in Fig. 16. It is evident that the combinations described in the previous article do not allow of any regulation of speed, the motor being simply shunt-wound with an unchangeable field. The purpose of the arrangement now to be described is to give speed regulation, which is accomplished by a changeable resistance in the field. The controller is mechanically operated.

As seen in Fig. 14, there are two cams I and II operating the armature and field-resistance arms A_a and A_f, respectively. Both arms are provided with dashpots D_a and D_f. Two more cams III and IV, shown in Fig. 15, operate the reversing switch, or pole changer, P; the one cam is intended to throw the switch for going up and the other for going down. While not visible in the illustrations, there

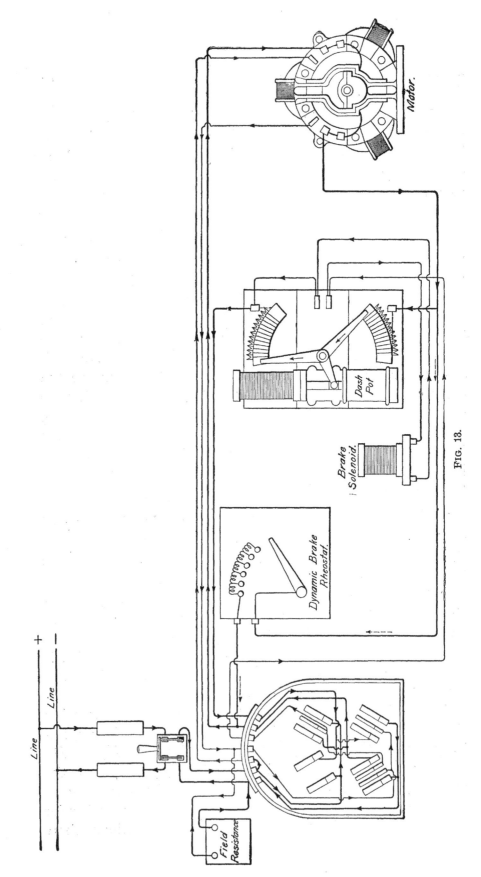

FIG. 13.

are other cams that operate various knife switches. All these cams are mounted on the shipper-sheave shaft S. The brake is the same as in the previous design.

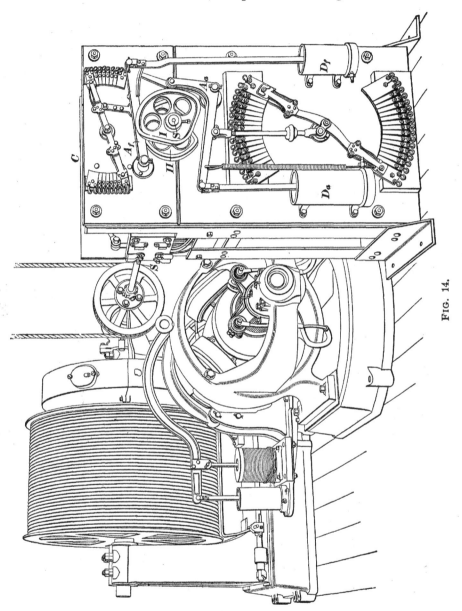

Fig. 14.

34. Fig. 16 is a diagram of the connections for this controller. (*a*) shows the external connections between motor, brake, and connection board B; (*b*) gives the internal

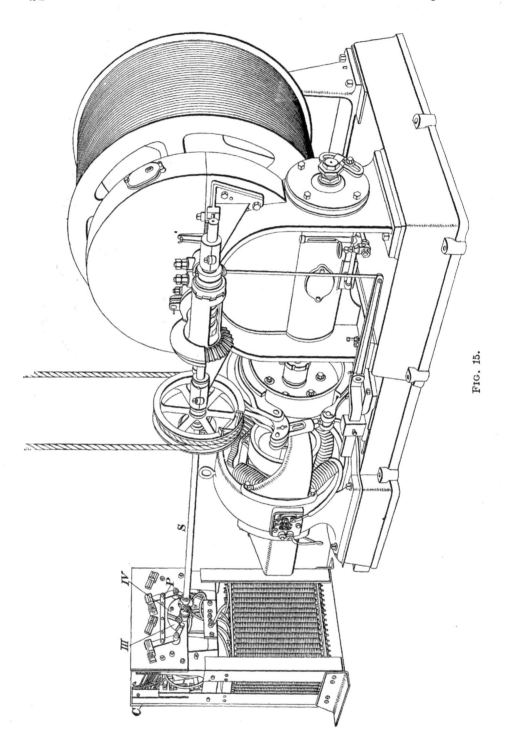

Fig. 15.

connections between the connection board B and the various clips and resistance blocks inside the controller. By swing-

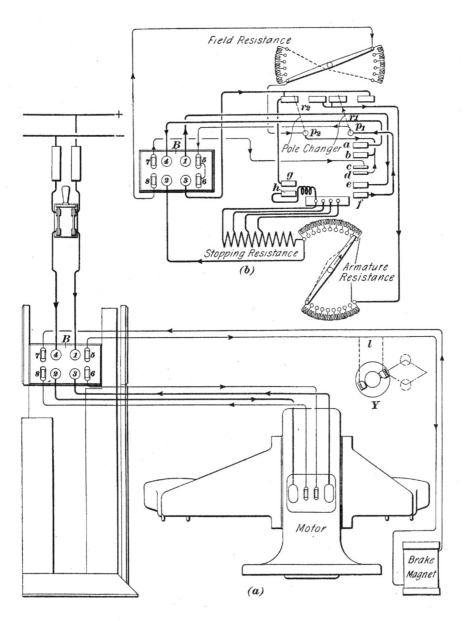

Fig. 16.

ing the shipper sheave to the right or left, switch blades connect the clips a and b, c and d, and e and f, completing the circuits. Thus, supposing the current to enter the system from the line at the binding post *1*, it goes to the clip e, over a blade or knife to the clip f, thence to the pivot p_1 of the pole changer, where it divides. One branch goes through the pole-changer arm r_1 and the armature resistance to binding post *2*, thence through the armature back to the binding post *3*, thence through the other pole-changer arm r_2 to the pole-changer pivot p_2, to the clip b, over the knife to the clip a, thence to the binding post *4*, and back to the line, thus completing the armature circuit. The other branch of the circuit goes from p_1 to the binding posts *5* and *6*, which, in turn, are connected, respectively, to the brake-magnet circuit and the shunt-field magnet circuit. The other terminal of the brake-magnet circuit is connected to the binding post *7*, whence the current flows over clips c and d, and b and a to the binding post *4*, and back to the line. The other terminal of the field circuit is connected to the binding post *8*, whence the current flows through the field resistance to p_2, b, a, post *4*, and back to the line.

35. The cam *I*, Fig. 14, on the shipper-sheave shaft is so arranged that after the circuits are closed the armature-resistance arm A_a is free to move, which it does slowly under the retarding influence of the dashpot D_a, gradually cutting out resistance until at the normal speed of the motor all resistance is cut out. After turning the shipper sheave a little farther, the cam *II* controlling the field-resistance arm A_f is released, but is retarded by the dashpot D_f. Thus the field resistance is slowly *cut in*, weakening the field and speeding up the motor.

36. Another pair of clips g and h, Fig. 16, is so connected that when a switch blade is thrown across them, the armature is short-circuited through the stopping resistance. This switch $g\,h$ is closed and the armature short-circuited when the other circuits are opened.

37. Motor Safeties.—The usual motor safeties, viz., limit stops and slack-cable safety, such as we have met in connection with belt and steam elevators, are used in the Elektron elevators. Their arrangement is shown in Fig. 15.

38. Another motor safety used is in the shape of a switch controlled by a centrifugal governor running in unison with the car, and which opens a switch in the brake circuit when the car attains undue speed. This safety is indicated at Y in Figs. 11 and 16, and is connected in series with the brake solenoid by opening the solenoid circuit at point l and inserting switch Y, as indicated by the dotted lines.

SEE ELECTRIC ELEVATORS.

39. Motors.—Fig. 17 shows one of the standard machines built by the A. B. See Manufacturing Company. A bipolar, drum-armature, compound-wound motor is used.

40. Transmitting Devices.—Among the transmitting devices, the step bearing shown in Fig. 18 is of peculiar construction. Both steps, that for the up trip and that for the down trip, are located at the free end of the worm-shaft and are easily accessible. The one is adjustable by means of the plug P in the cap C, while the other is made so by means of the nut N on the threaded free end of the shaft. The other end of the worm-shaft passes through a stuffingbox S, as in other machines. The worm and lower part of the worm-wheel are constantly running in oil.

41. Controller.—The controller, as shown in Fig. 17, is placed on top of the motor and consists of a box with three compartments, one of which is accessible from doors O, and another one from similar doors on the opposite side. The first, shown open in Fig. 19, contains the main reversing switch M and three snap switches N, N', and U, the blades, or knives, of the latter being mounted on the same lever, but insulated from one another. The switches are operated by a bar B, which, in turn, is linked to the rack R, Fig. 17, and operated by a pinion P fastened to the shipper sheave S. The opposite compartment contains a solenoid dashpot, a resistance lever, and resistance contacts very much the same as those shown in Figs. 7 and 10. The third compartment is located between the first-named two and contains resistance coils of German-silver wire. The walls of the compartments

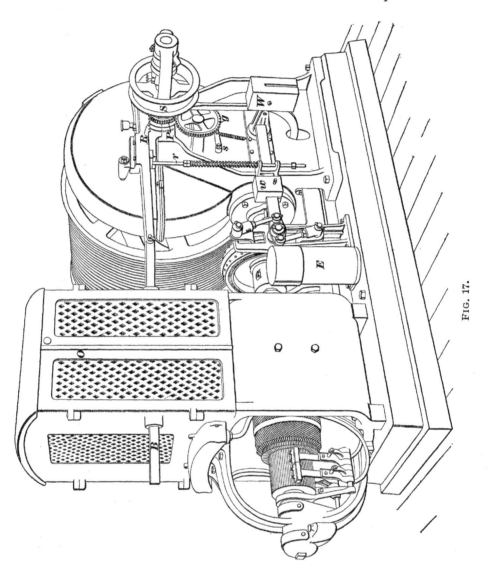

Fig. 17.

are cut away wherever they are not needed for the support of contacts or mechanisms, so as to give ventilation to the resistance coils; the doors and sides of the controller are perforated, as shown in Fig. 17, for the same purpose.

§ 38 ELEVATORS. 29

42. Brake.—The brake used in this machine is controlled mechanically and electrically. A spring-cushioned push rod r, Fig. 17, is operated by an arm fastened to the shipper sheave and forces the brake lever down to apply the

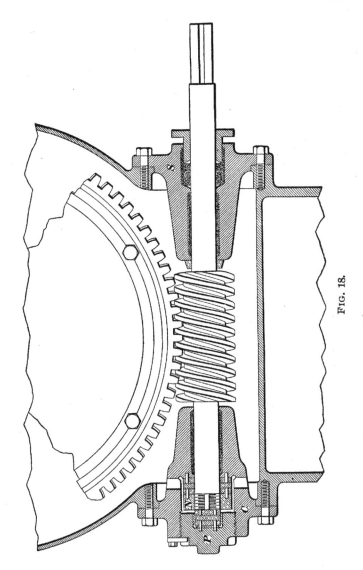

FIG. 18.

brake. A solenoid E holds off the brake as long as there is current in the armature with which the solenoid is connected in series. A weight W applies the brake, when the current is broken. There is also a dynamic-braking effect, the armature being short-circuited through resistance when current is shut off from the machine.

43. Motor Safeties.—This machine is particularly well provided with motor safeties. Not only the usual traveling-nut, limit-stop, and clutch-operating slack-cable safety are provided, but an extra limit switch is also provided, which breaks the current through the armature and brake solenoid

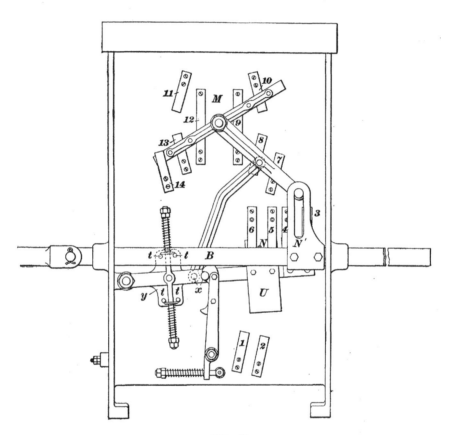

Fig. 19.

at the limits of car travel. This switch s, Fig. 17, located on the worm-gear casing below the drum shaft, is spring actuated and tripped by a stop on a gear g, which is one of a suitable train of gears driven from the drum shaft. The weight W throws in the clutch that connects the drum shaft with the shipper sheave when it is tripped by slack cable,

44. Electrical Connections.—Fig. 20 is a diagram of the electrical connections. The contact pieces are marked in the diagram the same as in Fig. 19. The circuits for the position of the controller shown in this figure are as follows.

45. *Armature Circuit.*—In the armature circuit the current passes through the + line to clip *3*; from clip *3* to clip *4* over the blade of the switch; from clip *4* to clip *19*; and from clip *19* to clip *18* over the switch blade, which is open only when the car overtravels the normal limits of travel; from clip *18* the current passes through the series coil of the electric brake to clip *5* and then to clip *6* over the switch blade; from clip *6* it passes through the series field of the motor, through the armature resistance and series coils on the armature resistance solenoid to clip *10* and then to clip *9* through the switch blade; from clip *9* it passes through the armature to clip *12*; from clip *12* to clips *13* and *14*; from clip *14* to clip *7*; and from clip *7* to the − side of the line.

46. In the armature circuit, when the pole changer is reversed from the position shown in Fig. 19, the current passes from the + line to clip *3* and then to clip *4*; from clip *4* to clip *19* and then to clip *18*; from clip *18* to clip *5* and to clip *6*; from clip *6* to clip *10* and on to clip *11*; from clip *11* to clip *12* and then through the armature, in a reverse direction, to clip *9* and then to clip *8*; from clip *8* to clip *7* and then to the − side of the line.

47. *Dynamic-Brake Circuit.*—When the controller is in its neutral position, that is, when the current is shut off from the machine, clips *1* and *2* are bridged by the switch blade U and the motor is short-circuited through the resistance, passing from clip *1* through the armature and then through the short-circuit resistance a to clip *2*.

48. *Electric Brake.*—The shunt coil of the electric brake obtains its current from clip *17*, clips *17*, *18*, and *19* being bridged by one switch blade, which is operated by the stop motion mentioned in Art. **43** and which stop motion

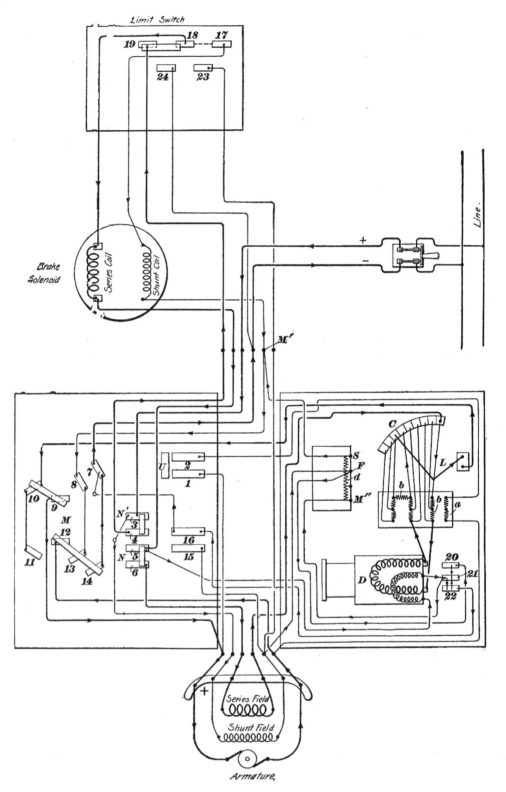

FIG. 20.

automatically breaks connection between clips *17*, *18*, and *19* when the car overtravels its normal limits. This switch is essentially an automatic safety switch, for it not only breaks the line current before it passes through the armature, but also breaks the current flowing through the shunt coils of the brake solenoid.

The $+$ side of the shunt coil is connected to the separate clip *17* instead of to the clip *18* in order that upon breaking the circuit the armature circuit may be disconnected from the electric-brake circuit, thus allowing the brake to act at once. Otherwise, the motor still running would send enough current through the shunt coil of the brake solenoid to keep it energized and thus prevent its action. The electric-brake circuit is, therefore, from clip *17* through the shunt coil to the terminal M', and from M' to clip *8* or *13*; from clip *13* to clip *14*, or from clip *8* to clip *7*, and thence to the $-$ side of the line.

49. *Path of Current in Starting Box.*—The shunt coil of the solenoid D, Fig. 20, gets its current from clip *5*; and after the current passes through the coil it enters clip *21*. The switch blade, or knife, that bridges clips *20*, *21*, *22* is drawn out of contact with the clips when the plunger of the solenoid reaches the end of its travel, when all the resistance in the armature circuit is thus cut out. Before the switch blade is removed, the current crosses on it to clip *22*; from clip *22* it passes to clip *16*, and thence to clip *7*, whence it goes to the $-$ side of the line.

When the contact is broken, the current is forced to pass from clip *21* to and through the resistance d from the terminal M'' to the terminal S; from the terminal S it passes to the terminal M', then to clip *8*, and so on to the negative side of the line. The resistance d is introduced in this circuit for the purpose of reducing the heating in the shunt coil and to reduce the current consumption after the solenoid has done its maximum work.

50. *Field.*—The field circuit of the motor is as follows: The current passes from the $+$ line to clip *3* and thence through the field to the terminal F of the resistance d. From F it passes to clip *20* and thence to clip *22*, to clip *16*, to clip *7*, to the negative side of the line. When the armature resistance is all cut out, contact between clips *20*, *21*, and *22* is broken, and the current is forced to pass through the portion of the resistance between the terminals F and S to the negative side of the line, provided the parallel connections from clip *15* to clip *16*, or at the limit switch from clip *23* to clip *24*, are broken. This resistance weakens the field on the motor and causes it to run at a higher speed. The contact between clips *23* and *24* is automatically made and broken when the car gets within about a floor from the top or bottom of its travel, and by the same stop motion that operates the limit switch. When the switch blade connects clips *23* and *24*, the resistance in the field is short-circuited; the field strengthens and the motor slows down. The switch blade bridging clips *15* and *16*, although situated at the machine, is withdrawn directly by the operator in the car during the last few inches of travel of his controlling lever, and he is thus enabled to weaken the field on the motor and run at a higher speed, but only after the car passes the first floor from the top or bottom and after all the resistance is cut out of the armature circuit.

51. In some of the See machines, the field is broken every time the motor stops. Fig. 20 is a diagram of a machine where the field is *on* all the time. Whether the field is to be left on or off is determined by the duty of the elevator. When the high-speed attachment is left off, a change in connections from those shown in Fig. 20 is made, Fig. 20 being a diagram of connections for a high-speed elevator running 250 feet per minute and over.

OTIS ELECTRIC ELEVATORS.

52. Motor.—The Otis Elevator Company makes a number of styles of electric elevators. They are all of the drum type, but have various kinds of controlling devices. Figs. 21

and 22 illustrate what may be termed the standard type of Otis elevators.

The motor used is the Eickemeyer bipolar, drum-armature, compound-wound type, the series coils of the field being

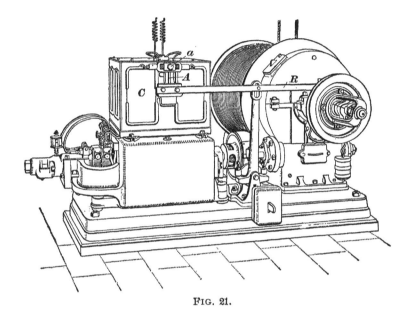

FIG. 21.

cut out after the starting resistance has all been cut out, that is, when the motor has acquired normal speed. This is done both on the up and down trip of the car.

53. Transmitting Devices.—With regard to the transmitting devices, it may be mentioned that either single or double worm-gearing is used, the latter for the larger sizes generally. In connection with the single worm a peculiar kind of step bearing is used. The purpose of this arrangement, shown in Fig. 23, is to increase the bearing surface, without enlarging the diameter of the step, by dividing the pressure between two surfaces, viz., the end surface s of the shaft and the ring-shaped surface s' of the bushing B. Now, it is well known to any mechanic that it is next to impossible to make the wear equal on two such separate surfaces unless special provision is made for it. This provision consists in this case of a couple of small levers l, l having three

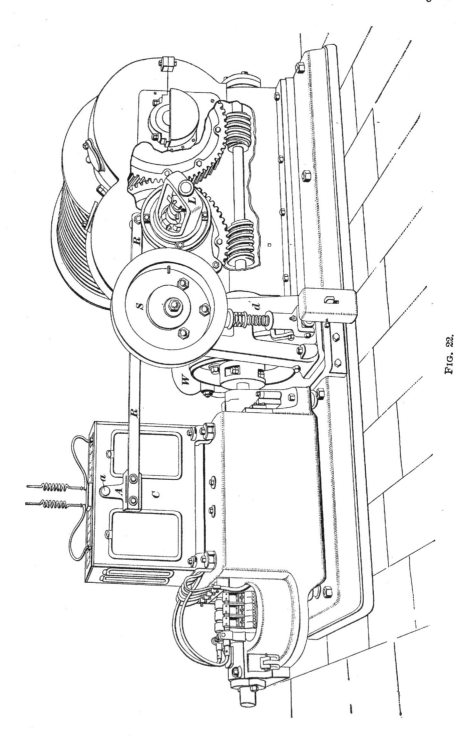

FIG. 22.

points each. One of these points, in the middle of one side of the lever, rests against an adjusting screw S, which is provided for the purpose with a circular groove. Of the other two points on the ends of the other side of the levers, one rests on the step plate P and the other on the bushing B. If the bushing wears faster at s' than the step plate wears at s, the shaft will move to the right, which will cause the levers to press on the bushing, and vice versa. Thus, the pressure is equally distributed over both surfaces s and s'. The screw S serves to take up the wear. The little equalizing levers l, l are held in place by being placed in slots in the sleeve or bushing B, and by a pin p that fits into semi-cylindrical grooves in the end of the levers. Buffers between the worm-gear and drum are used on all Otis electric elevators to absorb vibration.

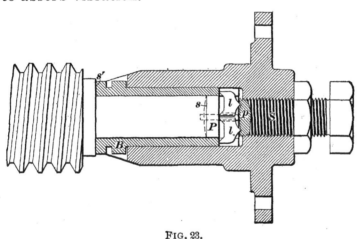

FIG. 23.

54. Controlling Devices.—The controller of the Otis elevators is box-shaped and is usually mounted on top of the motor, as shown at C in Figs. 21 and 22. It is operated by a rod R attached to the shipper sheave, which rod has an arm A on the other end, which engages by means of a part a with another arm or crank, hidden underneath the arm A; this crank is fastened to a shaft that reaches inside the controller box. In Fig. 24, which is a drawing showing the interior mechanism of the controller, this shaft is marked s. For **clearness, the two parts** (b) **and** (c) **of the mechanism are**

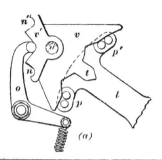

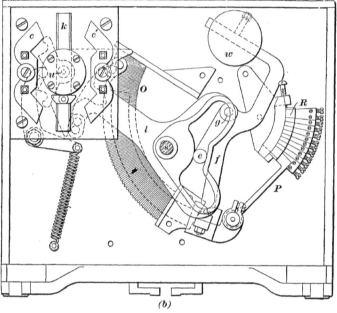

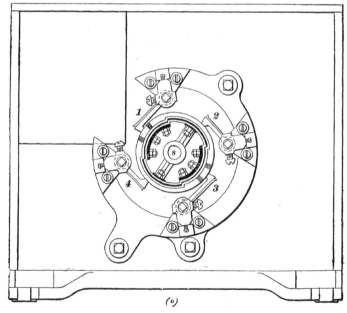

Fig. 24.

shown apart, while in reality part (*c*) is in front of part (*b*). Fig. 24 (*a*) is a detail view of some of the parts not very clearly shown in (*b*), where they are shown in dotted lines. The following is a description of the mechanism: the portion (*c*) contains the **reversing drum** mounted on and keyed to the shaft *s*; it has four contact plates insulated from one another. On these contact plates, of which two are long and two short, there rest four brushes *1*, *2*, *3*, and *4*, 90° apart. By turning the drum to the left, brushes *1*, *2* and *3*, *4* are made to rest on the same long contacts; while by turning the drum to the right, brushes *1*, *4* and *2*, *3* are brought into connection. The brushes are so connected to the armature and line that by turning the drum as aforesaid, the current in the armature is reversed. This will be plain from the diagram of connections given in Fig. 25. Behind the drum there is also fastened to the shaft *s* a lever *l*, Fig. 24 (*a*) and (*b*), carrying pins *p* and *p'*, which, when the shaft *s* is turned, engage a tooth *t* formed on a plate *v* pivoted at *u*. The plate *v* carries another plate *v'* having notches into which falls the end of a spring-actuated bell-crank lever *o*. By turning the shaft *s*, the plates *v* and *v'* are first turned around *u* until the end of the lever *o* rides on one of the sharp corners of the plate *v'*, whereby the spring of lever *o* is stretched. Turning the shaft *s* a little farther makes the end of the lever engage the inclined planes *n* or *n'*, which are so located that the spring causes the plate *v'* to make an additional quick rotary motion.

On the pivot *u* is fastened the blade *k* of the knife switch shown in the upper left-hand corner of Fig. 24 (*b*), and the quick rotary motion of the plate *v'* causes this blade *k* to snap between the clips *c* and *c'* of the switch. It is evident that on returning the mechanism to its middle position, the same snap action is caused by the two middle inclined planes of the plate *v'*, so that the switch blade *k* is quickly withdrawn from the clips *c*, *c'*, thus avoiding the formation of arcs; this is really the main object of the snap switch.

The other end of the lever *l* is formed into a cam of peculiar shape, which engages a pin *e* of a double-armed lever *f* pivoted at g. This lever f has fastened to its lower end a curved magnet core entering a solenoid O, as well as a contact arm P arranged to slide over resistance contact blocks R. The greater part of the weight of the magnet core and arm P is counterbalanced by a weight w on the other arm of the lever f, so that when free to move, the magnet core, while having the tendency of swinging out of the solenoid, will be pulled back into the same as soon as the current will produce enough magnetism to overcome the unbalanced weight of the core and the arm P. The lever f becomes free to move, however, only after the shaft s has been turned enough to make the circuit at the snap switch, the cam on the lever l holding all parts in position until then.

55. Supposing that the solenoid and the resistance R are in series with the armature, it will be seen that the operation of this apparatus is as follows: First the circuits are closed with all the resistance in the armature circuit and the motor starts up. By the time the motor has gained some speed the lever f is set free, and if the speed of the motor is such that the counter-electromotive force is enough to cut down the armature current to the desired amount, the solenoid will not hold the core, the latter will swing out, and the arm P sliding over the contact blocks R will gradually cut out the starting resistance. Should for any reason the armature current increase above the normal, the solenoid will pull back the core, throwing resistance into the armature circuit. It is thus seen that the solenoid performs two functions: first, that of cutting out the starting resistance; and second, that of a safety device. In stopping, the lever f is brought back into the original position by means of the cam on the lever l, making the arrangement ready for starting again.

56. The diagram of connections given in Fig. 25 will be readily understood. It is to be noticed that the series windings of the field are cut out after all starting resistance is cut out. A safety wire s connects the end of the solenoid with the first resistance contact. This wire will keep the

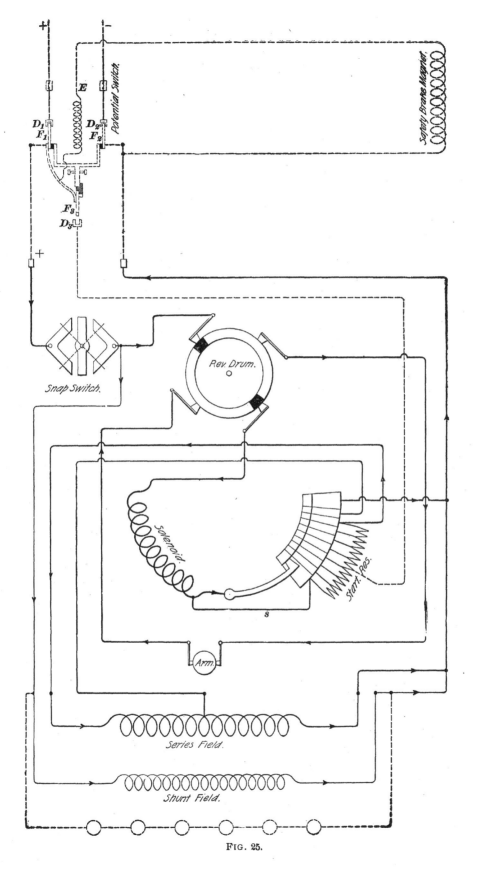

FIG. 25.

circuit closed even if, for some reason, the contact brush of the solenoid lever should fail to provide sufficient contact and thus stop the motor. The resistance coils are placed in a compartment of the controller box back of the mechanism shown in Fig. 24.

57. Brakes.—The brakes on the Otis elevators are of the band type. In the simpler forms, a steel band faced with leather encircles the pulley and is so connected to a weighted lever that the weight applies the brake. The lever is linked to the controller rod in such a manner that when the shipper sheave is turned either to the right or to the left the brake is released.

58. For high-speed service elevators, such as are shown in Fig. 22, a different kind of brake is used, for the reason that in such elevators the car must be stopped almost instantly without any possible slipping when the limits of travel are reached; while at any floor stop, midways of the travel, such instant stoppage is not so essential. The brake is, therefore, so arranged that it will be set in action by the limit stop much quicker and more effectively than by the ordinary device. The arrangement is shown in detail in Fig. 26.

On a stand A is a bearing a in which a short shaft s can revolve. To this shaft is keyed a crank-arm C, which in turn is connected by a rod R to the yoke of the limit-stop device L, Fig. 22. On the shaft s there is also keyed an eccentric E carrying another eccentric E'; the strap D encircling this outer eccentric is connected by a spring-cushioned rod d to the brake lever, and to it is also fastened the shipper sheave S, so that the latter, with the outer eccentric, turns upon the inner eccentric as a pivot. The outer eccentric has an arm C', Fig. 26, connected to the controller crank by a rod R', Fig. 22. To stop the car at intermediate landings, the brake is applied by turning the shipper sheave into the position shown in the figure, the outer eccentric pressing down on the brake lever. When, however, the limit stop is set in action, the inner eccentric is turned, which, having a greater throw than the outer one, gives more pressure to the brake.

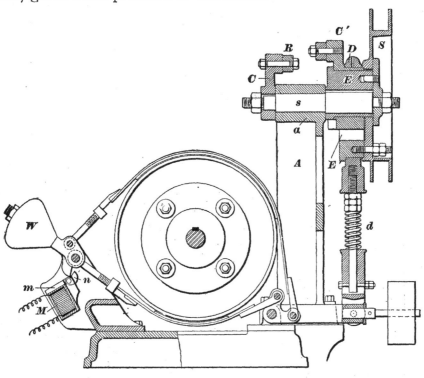

Fig. 26.

59. Another feature of the brake shown in Fig. 26 is the **safety magnet** M. This magnet serves to automatically apply the brake if the current should for any reason be interrupted in the system, and is placed in shunt with the motor, together with a so-called potential switch (of which we shall speak later), as shown in Fig. 25 in dotted lines. The armature m of this magnet has a projection, or nose, n which normally, that is, when a current of sufficient magnitude circulates through the magnet winding, holds in suspense a weight W connected to the free end of the brake band, as shown in Fig. 26. As soon as the current falls below the normal, the weight W trips the armature m and tightens the brake band. After the trouble causing this safety arrangement to act has been remedied, the weight is replaced into the position shown in Fig. 26

by operating the brake in the regular way. Dynamic braking is also resorted to.

60. Operating Devices.—For standard passenger and freight elevators, the simple hand rope is generally used; for high-speed elevators, hand-wheel devices or levers are preferred. To prevent accidental reversal of the motor in stopping, the tripping device (the lever l and the plate v) shown in Fig. 24 has considerable lost motion, or backlash.

61. Motor Safeties.—Besides the safety magnet brake above described, the usual limit-stop arrangement, consisting of yoke and traveling nut, and a clutch operating the slack-cable safety, the Otis Company generally installs with the magnet brake a so-called **potential switch.** This switch, shown in detail in Fig. 27, has three blades F_1, F_2, F_3, with

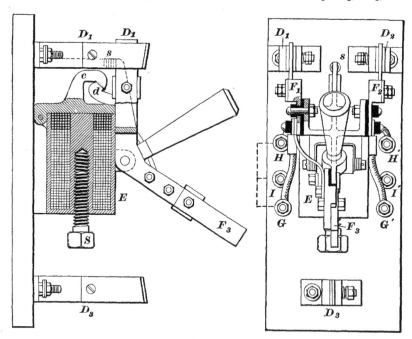

Fig. 27.

three corresponding double clips D_1, D_2, D_3, of which the first two are connected, as shown in Fig. 25, to the line wires, and the third D_3 to a wire leading to about the middle of the starting resistance. Blades F_1 and F_2 are connected to the motor circuit as shown, and F_3 to F_1. An electromagnet E placed in the shunt across the line in series with the safety-brake magnet holds the blades F_1 and F_2 in contact with the clips D_1, D_2 by means of a catch c on the armature of the magnet engaging a projection d on the fulcrumed lever carrying the blades. A spring s counteracts the magnet and causes the blades F_1, F_2 to leave clips D_1, D_2 and the blade F_3 to engage the clip D_3 when the current in the magnet windings falls below the normal. This has the effect of breaking the main circuit, releasing the safety brake, and thereby short-circuiting the armature through more or less of the starting resistance, according to the position of the resistance arm at the time. This short-circuiting acts as a brake on the motor, as is well known.

62. The usefulness of the potential switch extends beyond the use just explained. In Fig. 28, a method of connecting up the potential switch is shown, by which the potential switch not only performs its function in case of a fall of electric potential, but also in case of an undue increase of current in the line. For this purpose the switch magnet E, Fig. 27, has two windings with opposite magnetizing effect. One winding (the one next to the armature of the magnet) terminates in the binding posts H, H', while the other terminates in binding posts I, I'. These posts are, respectively, connected so as to throw the magnet winding HH' in series with the armature of the motor, and the coil II' in series with the safety magnet brake, as in the previous case.

The coils of the electromagnet are so proportioned that under normal conditions the shunt coil II' gives a stronger magnetic field than the series coil HH', and since they are wound in opposition to each other, the shunt coil will thus normally hold the switch closed. But if the potential in the line falls below the normal, the switch will be opened, the magnet not holding against the spring. Again, if the current in the armature circuit rises above the normal, the series coil of the magnet will produce a stronger field than normally, with the effect of weakening the field produced by the shunt coil, so that eventually the magnet will be demagnetized enough to let go of the switch lever. It is thus seen that the switch operates not only under a fall of potential

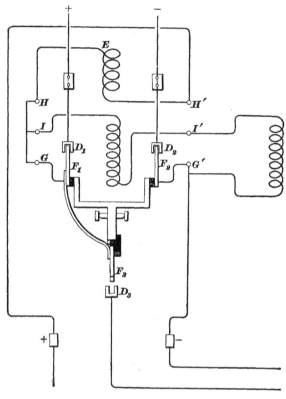

FIG. 28.

but also under an excess of current. The screw S shown in Fig. 27 serves to regulate the shunt field by screwing it in or out, decreasing or increasing, respectively, the resistance of the magnetic circuit of E.

ELEVATORS OPERATED BY ALTERNATING CURRENT.

63. While direct current is preferable for the operation of electric elevators, in many cases alternating current is the only source of power that is available. Two-phase or three-phase alternating current is generally used for elevator operation. Prior to the introduction of the two-phase and three-phase systems, alternating current was very little used for motive purposes because the single-phase alternating current motor would not start of its own accord under load; on the other hand, two-phase and three-phase motors give a good starting torque and will run up to speed in much the same way as a direct-current motor. An alternating-current induction motor consists of two main parts: the primary, or stator, which is the stationary part, and the secondary, or rotor, which is the revolving part.

The primary consists of a laminated body provided around its inner circumference with slots in which the primary coils are placed. These coils are connected together, and the terminals connect to the line when the motor is in operation. The secondary, or rotor, is also a laminated body provided with slots around its circumference in much the same way as a direct-current armature. In many induction motors, each of these slots contains a heavy copper bar, which is connected to a copper ring at each end of the armature, thus forming what is known as a squirrel-cage winding. In other types of machines, especially those that must give a good starting effort and are started and stopped frequently, the armature is provided with a three-phase winding and the three terminals brought out to collector rings mounted on the armature shaft. This is done so that resistance may be inserted in series with the armature windings when the motor is being started, and thus allow a good starting effort to be obtained without an excessive rush of current. In some cases resistance is inserted in series with the field, or stator, at starting instead of in series with the armature. This avoids the use of collector rings, but it does not give as good a starting effort for a given current as when the resistance is used in series with the armature. The student should note particularly that in the alternating-current induction motor no current is led into the armature from the line; in fact, there is no connection between the armature and the line. The armature currents are set up by the inductive action of the constantly shifting magnetic field that is set up by the two-phase or three-phase currents in the stationary field winding. This point should be borne in mind, as it will aid in understanding the connections to be described later.

OTIS ELECTRIC ELEVATOR WITH ALTERNATING-CURRENT MOTOR.

64. General Description.—Fig. 29 shows an Otis elevator operated by a three-phase induction motor A. This motor is of the type manufactured by the General Electric Company and is arranged so that a resistance is inserted in series with the armature windings at starting. In order to allow the insertion of this resistance, the armature is provided with three collector rings, shown at b, contact being made with the rings by means of carbon brushes. The motor operates the drum by means of a worm-gear, as already described in connection with other elevators. The starting, stopping, and reversing are controlled by a shipper sheave S operated from the car. When the shipper sheave is moved in either direction, a cam moves the rod r back and forth. In the figure the shipper sheave is in the neutral position. R is the reversing switch that connects the motor to the line and controls the direction of rotation of the motor. This switch is operated by the cam c. The shaft s of the switch carries a number of arms, which engage with suitable contacts when the switch is moved to either the up or down position. The cam c has three prongs that engage with prongs on a segmental gear G, and when the car reaches the limit of its travel in either direction, the traveling nut on the drum shaft causes G to open the circuit and stop the motion of the car. In the position shown in the figure, switch R is open; when thrown to the right, it makes connections for the car to go up, and when thrown to the left, it reverses the motor. Enough backlash is given between the prongs of the cam c and the lugs on the wheel G to insure safety against overthrowing the switch. When switch R is operated, the motor starts up with all the resistance in the armature circuit and means must be provided for cutting out this resistance as the motor comes up to speed. This is accomplished by the controller shown at O.

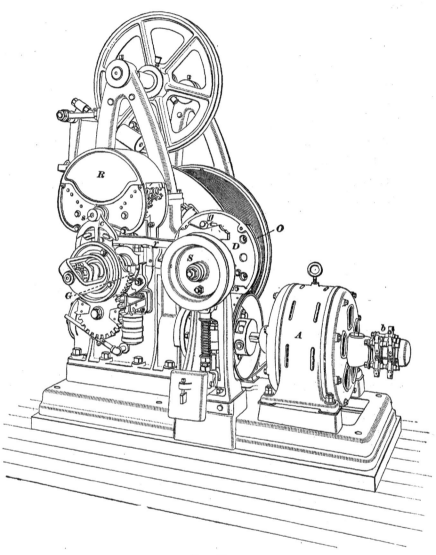

FIG. 29.

65. The Controller.—The controller is operated by means of the rod r, which raises the roller g whenever r is moved. The roller g is mounted on the end of a lever, as indicated in Fig. 30 (*b*). Fig. 30 (*a*) shows a rear view of the controller. D is the supporting cast-iron plate that carries the slate pieces S, on which are mounted a number of contacts l_1, l_1, l_2, l_2, etc. The hinged fingers f_1, f_1, f_2, f_2, etc. also carry contact pieces, and in the position shown in the figure, the fingers are in connection with their respective contacts mounted on S. When they are in this position, all the resistance is cut out and the motor runs at full speed, as will be shown later. As soon as r, Fig. 29, is moved, roller g is raised and casting E, Fig. 30 (*a*), is forced down, thus compressing the spring F and raising all the fingers. At the same time, the reversing switch is closed and the motor

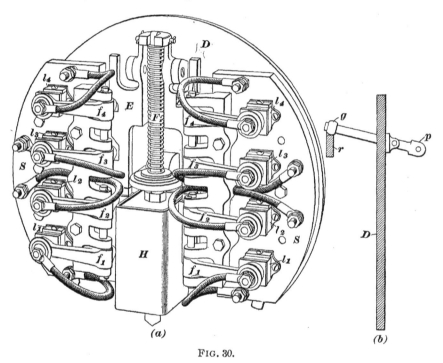

FIG. 30.

starts up with the resistance in. When roller g rides over the cam on r, the spring F forces up E, the upward motion being gradual because of the dashpot H. Casting E is provided with a number of cams, or notches, so placed that as E rises, the fingers f are closed down in pairs; i. e., the two lowest fingers first make connection with their contacts, then the next pair, and so on until all the contacts are closed, as shown in the figure. The closing of each of the pairs cuts out a section of resistance in each of two of the motor windings.

66.. Connections and Operation.—The operation of the reversing switch and controller will be understood by referring to Fig. 31, which gives the electrical connections. R is the reversing switch and M the main switch, which is operated by hand and is only used when the motor is to be cut off entirely from the line. Switch R is provided with six clips $4, 5, 6, 4', 5', 6'$, which engage with the blades or

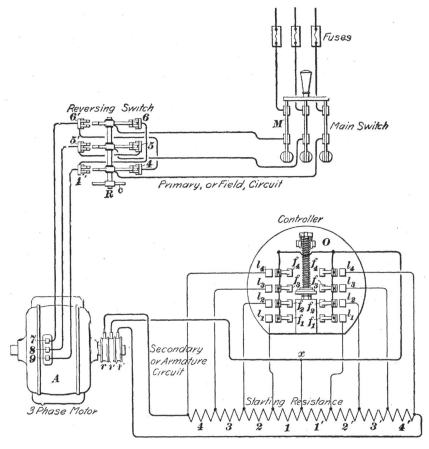

FIG. 31.

contact arms mounted on the shaft of the switch when the shaft is rocked by means of the cam c. In the position shown, the switch arms engage the right-hand clips and connection with the left-hand row is broken. The field terminals of the motor are $7, 8, 9$; and it is easily seen that when R is thrown over, the connections of 7 and 9 to the line are interchanged, thus reversing the motor. There is no resistance in this primary circuit, and the secondary or armature circuit in which the controller O is placed is entirely separate from the primary. In the position shown, all the fingers f_1, f_2, etc. are raised off the contacts l_1, l_2, this being the position they occupy at the moment of starting. The induced armature current in flowing from ring r' to r must take the path r'–x–1–2–3–4–r, thus passing through four sections of resistance. Also, in flowing from r' to r'' it must pass through the four resistance sections $1'$, $2'$, $3'$, $4'$. The insertion of this resistance in the armature windings keeps down the rush of current through the primary and results in a good starting effort. As the casting E, Fig. 30 (a), rises, fingers f_1 and contacts l_1 make connection, thus short-circuiting sections 1 and $1'$ of the resistance. As E rises still farther, sections 2 and $2'$ are cut out by f_2 and l_2 making contact, and so on, until all the fingers are down and all the resistance cut out. In passing from ring r' to r, the current now takes the path r'–f_4–l_4–r and there is no resistance in circuit. When, therefore, the fingers are all down, rings r, r', and r'' are connected together and the induced armature currents are provided with a closed circuit in which there is no resistance other than that of the copper armature conductors and the connecting wires.

67. The number of steps of resistance depends on the service to which the elevator is to be put. For example, some controllers are provided with only three sets of contact fingers, as it is found that three sections of resistance are sufficient to give a smooth start. The connections for a two-phase motor are practically the same as those shown, so that it is not necessary to describe them in detail.

ELECTRIC ELEVATORS WITH MAGNET CONTROL.

68. General Features of Magnet Control.—In most of the controlling devices so far described for electric elevators, the cutting out of the starting resistance is accomplished by means of an arm carrying a contact that slides over a series of plates, or contacts, connected to the sections of the resistance. This method works very well if the contact brush and contact plates are kept in good condition, but if either of them become rough or burned, the starting rheostat rapidly gets into very bad shape on account of the poor contact and consequent burning action. This is especially the case if the motor requires a large current for its operation, because the larger the current, the more perfect must be the connections made by the rheostat contacts, and a contact that is at all defective will very soon give rise to burning and cutting.

69. In order to avoid the use of a sliding contact with its accompanying contact plates, the so-called magnet system of control has been devised, in which the resistance is cut out by a series of electromagnetic switches, each one of which operates independently and which is so designed that it will handle a large current with very little burning or arcing. As these switches are simply of the make-and-break variety and have no sliding contacts, any small amount of burning that may take place does not interfere with the operation of the controlling outfit. There are many ways in which the system of magnet control may be applied. The electromagnetic switches may be arranged to operate automatically as the motor increases in speed; they may be controlled entirely by a controlling switch on the car, or part of them may be controlled automatically and part from the car. These resistance-controlling switches, together with the other electromagnetic switches necessary for closing the main circuit and reversing the armature connections, are mounted on a switchboard, which is usually separate from the elevator motor and hoisting mechanism.

70. Elementary System of Magnet Control.—Before taking up an elevator with magnet control, we shall consider the elementary arrangement shown in Fig. 32. This diagram is intended merely to illustrate the principle and does not represent any special controller. It shows an ordinary shunt motor M with its starting resistance R controlled by the two magnets S, S'. The starting and reversing switch is shown at A, and in this case it is supposed to be operated by hand. Of course, if the motor were used in connection with an elevator, switch A could be operated from the shipper sheave. When the switch is in the position shown, the motor runs in one direction, and when it is thrown over so that the blades occupy the position shown by the dotted lines, the motor is reversed. The starting resistance is divided into two sections a, b, which are successively

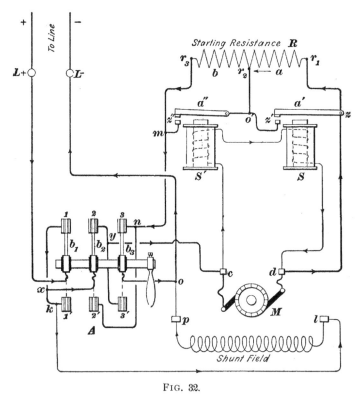

FIG. 32.

short-circuited by the electromagnetic switches S, S' when the motor comes up to speed. The windings of S, S' are connected in series across the armature terminals, forming a shunt circuit to the armature. When the main switch is closed, all the resistance is in series and the pressure across the armature terminals and coils S, S' is very small; consequently, very little current flows through S, S'. However, as the motor speeds up, its E. M. F. increases and the pressure across the brushes increases, and this increases the current through S, S'. The armatures of these switches are so adjusted that S will operate with a smaller current than S'; consequently, as M comes up to speed, S closes and cuts out section a of the resistance by short-circuiting it. As the speed increases still further, the current through S and S' becomes strong enough to operate S', and section b is short-circuited, thus connecting the motor armature directly to the line. Suppose that the motor is to be started and that switches A, S, and S' are in the positions shown. The path of the current through the shunt field is as follows: $L+$ $-b_1-$ $1-k-l-$ through shunt field$-p-o-L$. The path of the main current through the armature and starting resistance is $L+$ $-b_1-1-x-b_2-2-y-c-$ through armature of motor$-d-z-r_1-$ r_3-3-b_3-o-L. When the current through the shunt-magnet circuit $C-S'-S-d$ has become strong enough to pull down armature a', contact is made at z' and the main current on reaching z takes the path $z-a'-z'-o'-r_2-r_3-3-b_3-o-L$, thus flowing past section a of the resistance that is short-circuited. When S' operates, the current takes the path $z-a'-z'-o'-a''-z''$, and so on, the whole of the resistance being thus short-circuited. Any arcing, or burning, that may occur will take place at contacts z' and z'', and this can easily be taken care of by providing suitable contacts. Moreover, it will be noticed that the closing of an armature short-circuits the resistance, and that when an armature opens, the circuit is not broken, because the current still has the alternative path through the resistance. The result is that when the armature leaves its contacts there is but little sparking.

71. When the motor is to be run in the reverse direction, switch A is thrown over to the position indicated by the dotted lines. This does not change the direction of the current through the shunt field, but it reverses the current through the armature, the path being as follows: $L+$ $-b_1-$ $1'-x-b_2-2'-n-r_3-r_1-d-c-y-3'-b_3-o-L$. Since the current through the armature is reversed while that in the field remains the same, the direction of motion is reversed.

The scheme of using electromagnetic switches to control the starting resistance has been embodied in the controllers of a number of different manufacturers. It has been found that it is not necessary to provide a great many resistance sections and resistance-controlling switches in order to give a smooth start. The actual number needed depends, of course, on the conditions under which the motor is operated. With an ordinary sliding-contact rheostat, it is necessary to provide quite a large number of resistance sections, in order to keep the voltage between adjacent contact plates down to the small amount necessary to avoid sparking when the arm slides from plate to plate. With electromagnetic switches the number of sections can be much smaller, because this precaution is not necessary. Moreover, when the cutting out of resistance is controlled by switches that are in turn controlled by the counter E. M. F. of the motor, the resistance is never cut out until the armature has come up to such a speed that it is able to take care of the increased current. The resistance is, therefore, cut out just when the armature is ready for it and not before; such being the case, fewer resistance sections are necessary than if the cutting out were controlled by hand.

72. With most high-speed passenger elevators using this method, the switches that perform the same duties as A, Fig. 32, are operated by electromagnets or solenoids, thus doing away with the shipper sheave with its cable, cams, and other switch-operating devices and replacing them by an electric cable connecting the car-operating switch to the switchboard.

73. The car-operating switch replaces the ordinary operating wheel or lever used for operating by means of a cable. The cable running from the operating switch to the switchboard carries the wires that connect to the electromagnetic switches, and as these switches require only about $\frac{3}{4}$ ampere for their operation, the wires in the controlling cable do not need to be large. This method of control is being used quite largely for various kinds of service, and, as pointed out above, it has advantages over the older sliding-arm method of controlling resistance. In order to illustrate its application in practice, we will describe two controllers made by the Otis Elevator Company and covered by patents owned by them.

OTIS ELEVATOR WITH G. S. MAGNET CONTROLLER.

74. General Description of Elevator Machine.— Fig. 33 shows a direct-connected Otis electric elevator for use with magnet control. The motor M operates the drum D by means of double worm-gears. This particular machine is provided with back gearing between the motor shaft and worm-shaft, so that unusually heavy weights, such as safes, may be lifted. It will be noticed that there is no electric controller connected to the machine other than the brake magnet N' and the stop-motion switch M'. The brake magnet is a powerful solenoid that operates against the spring G, so that when the magnet is energized the band brake is released, and when current ceases to flow through the magnet, the brake at once goes on. The stop-motion switch M' will be described more in detail when the electrical connections are taken up. Its function is to cut off the current and stop the motor whenever the car approaches the limit of its travel in either direction. Under ordinary running conditions, the intermittent gear g remains in the central position shown in the figure. When the car approaches the limit of its travel, the safety nut on the shaft of the worm-gear causes a pin to engage with g, thus making it swing over. This operates a switch arm inside the casing M', which breaks electrical connections and slows down the motor. When the safety nut makes another revolution, g is swung over another notch and the motor is stopped completely. The mechanical features of the hoisting machine are similar to those that have already been described and do not call for special attention.

75. General Description of Otis G. S. Magnet Controller.—Fig. 34 is a general view of the Otis G. S. magnet controller. The controlling devices are mounted on a heavy

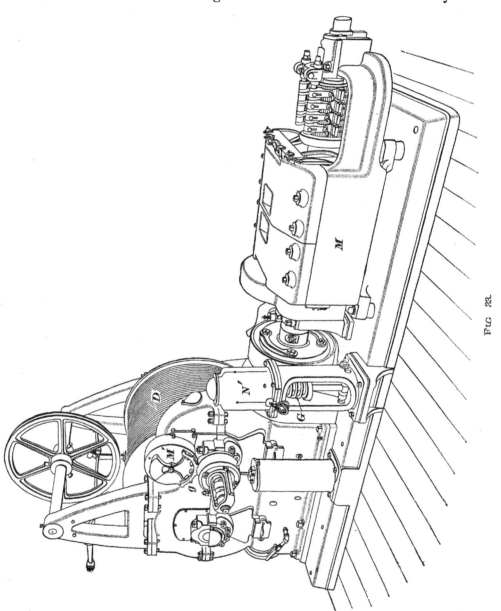

Fig. 34.

slate panel A, which is in this case supported on an iron framework B that also serves to house the resistance coils. With many controllers, the resistance is placed in a case arranged behind the switchboard. The various electromagnetic switches necessary for controlling the direction of motion of the car and the cutting out of the starting resistance are mounted on A.

FIG. 34.

In Fig. 34, B' is the potential switch, the use of which has already been explained. It is a protective device and is not concerned with the regular starting and stopping of the elevator. When the elevator is in operation it remains closed. Switches C', D', and E' control the main current.

Switch H' controls the brake and the two groups of switches G' and F' control the resistance. The group of four switches F' controls the starting resistance and the pair of switches G' controls the stopping resistance. With these controllers the motor is stopped by allowing it to act as a generator, thus providing a dynamic-braking action in addition to that of the band brake. In order to allow a smooth braking action, the current generated by the motor is passed through a resistance, and this resistance is cut out or in by magnets G'. The main operating magnets C', D', and E' are of the solenoid type, and when they are not excited the plungers are down and the upper switch contacts, as c, for example, are separated from the fixed contacts d. The movable contacts c are mounted on rocker-arms a, a' pivoted as shown at b. The plungers of the two switches D' and E' are connected by a lever l, as shown, so that when one contact lever a is up, i. e., the upper terminals in contact, the other lever a' is down, and it is impossible for both levers to occupy the up or down position at the same time. The operation of these switches will be understood more clearly by referring to Fig. 35 (a). Switches F' and G' are arranged as shown in (b) and switch H' is as shown in (c). These sketches are intended merely to indicate the operation of the switches, so that the diagram of connections to be given later may be readily understood; hence, particular attention has not been paid to the mechanical details. In (a), when the magnet draws up the plunger, lever a is moved so that c and d make contact, and contacts c' and d' are, of course, opened. Contacts d, d' are graphite blocks mounted on spring holders, the object of the graphite being to prevent damage from burning or sparking, and especially to obviate the danger of the contacts fusing, or sticking, together as might possibly occur if both were of copper. Fig. 35 (b) shows the construction of the resistance-controlling switches; f is the magnetizing coil, which is made to serve for the whole group of switches by embracing the series of cores h, as indicated at F', Fig. 34. G is the magnet casting that carries the

§ 38   ELEVATORS.   61

series of cores h, opposite each of which is hinged the armature a carrying an insulated contact c, which makes contact with d when the armature is drawn down. When current flows around coil f, all the cores are magnetized to about the same degree, but the armatures are not all attracted because they are adjusted to different distances from the pole pieces s by means of adjusting screws p that rest against lugs r. The armature with the shortest air gap between a and s is first attracted, then the next, and so on, the armatures closing in succession as the magnet increases in strength on account of the motor speeding up. The resistance is thus automatically cut out by steps, as explained in connection with Fig. 32.

Fig. 35 (c) shows the switch indicated by H' in Fig. 34. It is practically the same as (b), except that it is provided

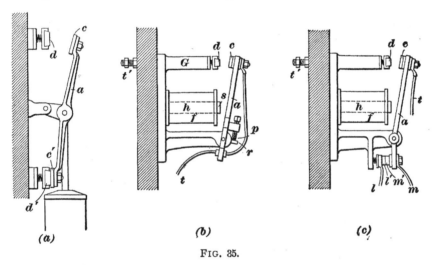

FIG. 35.

with two insulated back contacts l', m' to which the leads $l\ m$ are connected. When armature a is unattracted, l' and m' are in contact; when a is attracted, contact between l' and m' is broken and contact between c and d is closed. The switch contacts that are most liable to arcing are provided with magnetic blow-out coils. These are coils provided with an iron core so placed that a magnetic field is set up between the contacts, and as soon as the arc forms, it is forced across the field and broken almost instantaneously.

76. Car-Operating Switch.—Fig. 36 shows the style of car-operating switch used with the magnet controller. When the motor is stopped, the handle occupies the vertical position and is thrown to the left or right, according as the car is to go up or down. When the cover is closed and the switch in use, sliding contacts c, c bear against the arcs a, a, b, b; when the switch is off, they bear against the insulating pieces d, d. The contacts on the back of the operating lever press against segments e, thus making the required connection. By adopting the construction shown, no current flows through the hinge f. The exact arrangement of the contact segments varies with different controllers, as the starting and running requirements are not always the same for different installations. The operating switch for which the connections are shown in Fig. 38 is somewhat simpler than that shown in Fig. 36, and requires fewer wires and contact arcs, but its general construction is the same.

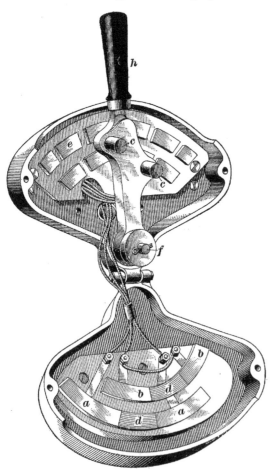

FIG. 36.

77. Stop-Motion Switch.—Fig. 37 shows two views of the stop-motion switch shown at M', Fig. 33. The use of this switch has already been explained, but Fig. 37 shows the construction and will aid in understanding the electrical connections. The arm a, which is operated by the intermittent segmental gear g, normally occupies the horizontal position, but is swung around whenever the car reaches the limit of its travel. Contact brushes are mounted on the arm, and these rub on the contact arcs b, b. When the arm is swung

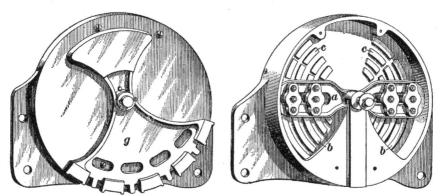

FIG. 37.

around in either direction, one set of brushes leaves the long contacts b, b and passes on to the short pieces c, c. Arcs c, c are, with the controller to be described, not connected to anything, but serve as bearing pieces; as soon, therefore, as the contact arm slides on to them, electrical connections are broken, which causes the motor to stop.

78. Connections for G. S. Controller.—Fig. 38 shows the general scheme of connections for the G. S. controller. In order to simplify the diagram, the relative positions of a few of the parts have been changed; for example, the starting resistance and the extra-field resistance $r\ r'$ are shown connected directly to their switch contacts. The relation of the various switches is the same as shown in Fig. 34, except that their order is reversed, because all the connections are made on the back of the board; corresponding switches in Figs. 34 and 38 are lettered alike. In order to facilitate the tracing out of connections, all *fixed* contact pieces on the switches have been shaded, while all *movable* pieces have been left open. For example, on switch E' the shaded contact pieces $5'$, $6'$, $7'$, and $8'$ are mounted on the slate panel and the open contacts *5, 6, 7*, and *8* are mounted on the tilting arm shown in Fig. 35 (*a*). Also, contacts that touch each other are marked with similar figures, i. e., when switch E' is pushed up, *5* makes contact with *5'* and *6* with *6'*. A few details, such as blow-out coils, have been omitted, as they are not necessary to illustrate the operation of the controller. The car-operating switch controls switches E', D', C', and H'; switches M', F', and G' operate automatically. The main switches E' and D' are each provided with two coils. One of these coils is of fine wire, and the current in it is controlled by the car-operating switch. The lower coils are of coarse wire, and carry the main motor current; these coils are arranged below the fine-wire coils and, when energized, hold the switch down. When switch C' operates, *13* makes contact with *13'*, *14* with *14'*, and contact between *15* and *15'* is broken. When H' operates, contact is made between *16* and *16'* and broken between *17* and *17'*. When switches G' and F' operate, contact is made between *1* and *1'*, *2* and *2'*, etc. The movable contact pieces of the stop-motion switch M' occupy the horizontal position shown until the car reaches the limit of its travel in either direction. The full-black segments on this switch are not connected to anything, being in this case bearing surfaces only. Switch P' is provided with a pair of contacts on each side of the "off" position. Two contacts u' and d' are longer than the others fu and fd, so that the lever makes contact with the former before the latter. To avoid confusion, a wire is shown connected to the lever instead of carrying the current to it through sliding contacts, as is done on the switch shown in Fig. 36. An additional safety switch S is sometimes provided to stop the elevator in emergencies, but it is not in use under normal conditions.

79. Type of Motor Used With G. S. Controller. Before taking up the action of the controller, it will be well to consider briefly the type of motor used with this system of control. In order to get the elevator under way quickly, it is necessary that the motor should give a strong starting

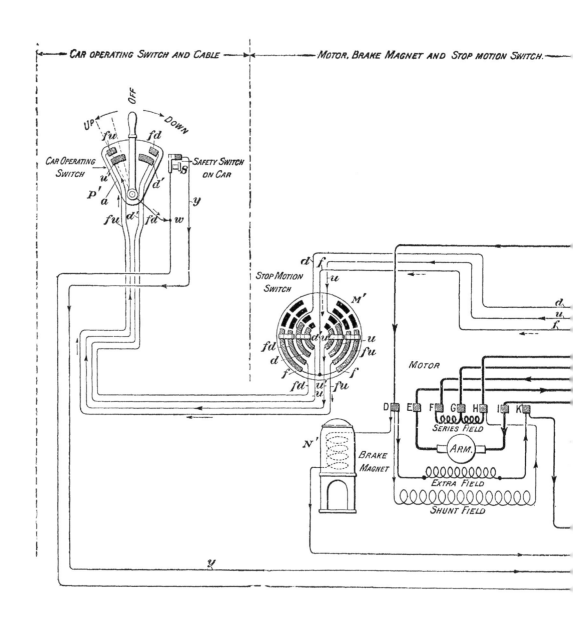

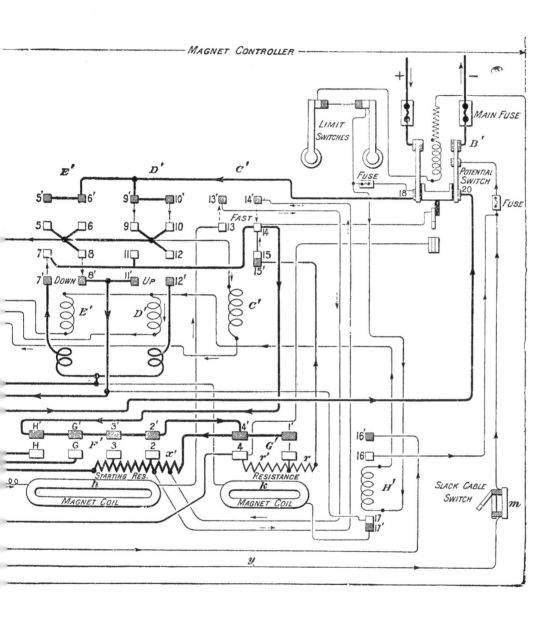

torque. This is provided for by the series field. The shunt field furnishes the excitation after the motor has attained its speed. In addition to these two windings, a third, or extra-field, winding is provided. This winding aids in providing a field when the motor is being brought to a stop, by allowing it to act as a dynamo; it also aids to some extent in providing a strong field at starting. It should be remembered that a shunt-wound motor will run as a generator if it is disconnected from the mains when up to speed and a path provided between the brushes for it to send a current through; it is not necessary to reverse either field or armature connections in order to make it generate, as is the case with a series motor.

80. Operation of Controller on First Point.—Suppose that the car is to be run up and that the lever of P' is moved to the left until the arm comes in contact with the long arc u', but does not touch the contact fu. The operating current then flows as follows, starting from point *18* on the + side of the potential switch: *18*, through coil of switch H', through coil of the "up" magnet D', through wire $u\ u$ to stop-motion switch M', to contact strip u', by way of the horizontal strip, through flexible car-operating cable to contact u' on car-operating switch, through lever to w, thence through safety switch and wire $y\ y\ y$ to the negative side of the potential switch. This current operates switches D' and H'. Switch D' is drawn up by the fine-wire coil and contact is made between *9, 9'* and *10, 10'*, as indicated by the dotted lines. This allows current to flow through the shunt field by way of the path +–*18*–*9'*–*9*–*D*– through shunt field–H–*19*–*20*. The operation of H' releases the brake by connecting points *16* and *16'*, because point D connects through D' to the + side of the circuit, and when *16* and *16'* are in contact, the other terminal of the brake magnet connects to the negative side of the circuit. This releases the brake and allows the motor to start as soon as current flows through the armature. When switch H' operates, points *17, 17'* are separated so that no current can flow through coil *k*, and, hence, switches G' are open so long as the motor is working. As soon as switch D' is forced up, switch E' makes contact between 7, 7' and 8, 8', because of the connecting lever *l*, Fig. 34. The main current then takes the path indicated by the arrowheads as follows: *18–9', 10'–9, 10–8–8'–I*–through armature of motor to E–through series coil of switch E'–*7'–7–14– H'–G'–3'–2'–4'*–through whole of starting resistance to F– through whole of series field to *H–19–20* to negative side of the circuit. The current through the series coil of E' holds the switch arm down firmly. The motor, therefore, starts up with the two sections of the series field and all the starting resistance in series with the armature. The extra field is in series with the resistance $r\,r'$, and the two together are in shunt with the armature. This is easily seen by tracing the path through the series field, beginning at point D, as follows: D–through extra field–*K'–4–r'–r– 15'–15–14*–etc.

Fig. 39 represents, diagrammatically, the connections that are made on the first position of the switch P'. The

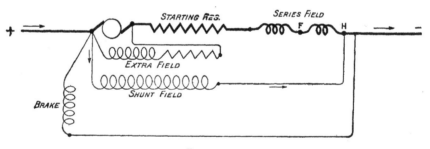

FIG. 39.

extra field does not supply nearly as much magnetizing power at starting as when the motor is stopping, because, at starting, the pressure across the armature terminals is small. On this point of the controller, therefore, the motor starts up, but would run the elevator at a slow speed because of the resistance in circuit. It should be noted that as long as the operating handle rests on u' or d', the resistance switches F' are inoperative, because one terminal of coil **h** connects to contact *13* on switch C', which is open.

81. Operation on "Fast Up" Position.—When the handle of P' is moved over farther, so as to make contact with the fu (fast up) contact, current flows through the solenoid of switch C', thus forcing the switch lever up and making contact between *13, 13'* and *14, 14'*. At the same time contact is broken between *15* and *15'*, thus opening the circuit through the extra field, which is now no longer needed, as the motor is by this time well under way. The operation of this switch allows current to flow through the magnet coil h and resistance x', as indicated by the dotted arrows, beginning at point *21*. This operates the group of switches F'. First *2* and *2'* are connected, thus cutting out the first section of resistance, then *3, 3'*, cutting out the second section, then G, G', cutting out what is left of the resistance and also one half of the series field; finally, H and H' are connected, thus cutting out the other half of the series field. The motor has now attained its maximum speed, and the path of the main current is $+-18-9', 10'-9, 10-8-8'-I$-through armature-$E-7'-7-14-H'-H-19$ to negative side of circuit. The motor now operates as a plain shunt machine with no resistance in circuit.

82. If the operating switch were moved in the reverse direction, switch E' would be moved up and switch D' would be down, as can be readily seen by tracing the connections. The main current then takes the path *18-6', 5'-6, 5-12-12'*-through series coil of $D'-E$-through armature-$I-11'-11-14$, and so on as before. The current in the armature flows in the reverse direction to what it did before, while the current in the fields remains unchanged; the car, therefore, moves down.

83. Action of Controller on Slowing Down and Stopping.—Suppose that the elevator is running on the "fast up" point and that the handle is moved back until it leaves contact fu, but still rests on contact u'. Switch C' will be opened, and this will cause switches F' to open, thus cutting the resistance and series field back into the circuit; the extra field will also be connected, because *15* and *15'* will make contact, the resistance r r' being in series with the extra field. Quite a large current will now flow through the extra field because the potential across the armature is high. When, therefore, the handle is moved back from the fast position, the field of the motor is greatly strengthened and all the resistance is cut back into circuit, thus rapidly lowering the speed of the motor. On account of the decrease in speed and the cutting in of the resistance, the pressure across the brushes is considerably decreased when the handle is moved from the fast position. When the operating handle is moved to the off position, switches D' and H' are opened. D' breaks connection with the line, and H' sets the brake by separating points *16*, *16'* and opening the circuit through the brake magnet. At the same time, points *17*, *17'* are brought into contact, thus connecting coil k across the armature terminals. The pressure across the armature terminals is large enough to cause switches *4*, *4'* and *1*, *1'* to close, thus cutting out r' r and connecting the extra field across the armature. The armature thus generates current, which takes the path *1–8'–8–D'*–extra field–K'–*4*–*4'*–*14*–*7*–*7'*–*E*. The current through the extra field remains in the same direction as it did when the motor was run from the line and hence assists in keeping the field magnetized and bringing the motor to a stop quicker than if the shunt field only were used. As the motor slows down, the magnetization supplied from the shunt-field coil diminishes; hence, the provision of the extra field supplied with the current that the motor furnishes when running as a generator greatly increases the braking action. The generating action soon slows the motor down, and as the pressure across the armature terminals decreases, switches *4*, *4'* and *1*, *1'* open in succession, because K is no longer able to hold them. This cuts resistances r', r back into circuit with the series field, thus making a smooth stop and leaving the resistances r, r' in series with the extra field ready for the next start. Of course, while this action is taking place, the band brake is also on because the dynamo-braking action decreases as the speed decreases, and hence would not answer, in itself, for bringing the motor to a full stop. All these actions take place in a very short space of time, but the effect is to stop the motor smoothly and quickly, and the car is at all times easily controlled by switch P', the cutting out of the resistance and the connections necessary to produce the dynamo-braking action being made automatically.

84. The stop-motion switch M' merely brings about automatically the same connections that P' should, in case the operator failed to move P' when the car reaches the limit of its travel. When the contact arm is swung around by the action of the traveling nut, as already explained, contact is first broken, if the car is ascending, between the $f\,u$ and f contact arcs, thus slowing down the motor; and as the car travels still farther, contact is broken between arcs u and u', thus applying the brake and stopping the car.

OTIS NO. 6 MAGNET CONTROLLER.

85. General Description of No. 6 Controller.— Fig. 40 shows the general arrangement of the Otis No. 6 magnet controller. This is a later type than the G. S. controller previously described, and although its mechanical details are quite different, its principle of operation is almost identical. The resistance, which is usually in the form of cast-iron grids for controllers of large capacity, is arranged behind the board and does not appear in the figure. The various switches are marked A', B', C', 1, 2, 3, 4, 5. Switches A', B', C', and 1 are operated by the car-controlling switch; the other switches operate automatically. Whenever a switch, for example C', operates, its plunger e is drawn up, thus bringing the copper disks d, d' up against the contact fingers f, f'. When a switch is deenergized, its plunger drops and the disks make contact with the lower fingers where any are provided. When a disk is drawn up, it first makes contact with the auxiliary carbon contacts x, and as it is pulled up still farther, it bears against a copper contact on a finger hinged at the same point as the finger that carries the carbon contact. When a disk drops, it first breaks contact at the copper surfaces, and the final break

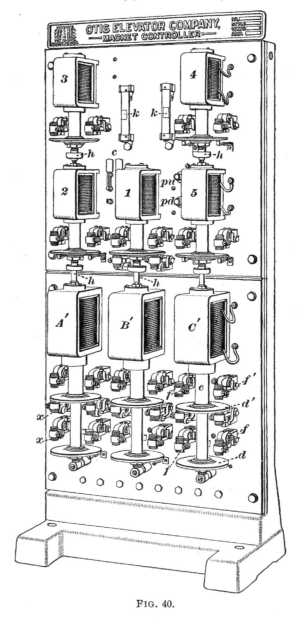

FIG. 40.

takes place between the copper and the carbon terminal, so that there is no danger of sticking. The carbon pieces are threaded, so as to prevent their working loose and sliding through the holders. The plungers with their contact plates are free to revolve, and the motion of the switch gradually works them around so that whatever burning takes place is spread around the whole disk instead of in one place only. Switches *2*, *3*, *4*, and *5* operate automatically, one after the other, and the voltage at which they operate is adjusted partly by regulating the initial position of the plunger by means of the adjustable stops h, and partly by inserting a resistance in series with each solenoid. The main fuses are shown at k, k; they are of the enclosed type. The small knife switches shown at c are used for cutting off the car-controlling switch, so that the motor cannot be started from the car. Push buttons $p\,u$ and $p\,d$ are used to allow the motor to be operated from the board. These devices are very useful when tests are being made to locate trouble, but under ordinary working conditions they are not in use. Switches A' and B' control the direction of motion of the car. When A' operates, the car descends, and when B' operates, it ascends. Switch C' closes and opens the main circuit.

86. No. 6 Car-Controlling Switch.—Fig. 41 shows the car-controlling switch used with the No. 6 controller, the cover being removed in order to show the working parts. The operating handle is shown at h, and it normally occupies the central, or off, position.

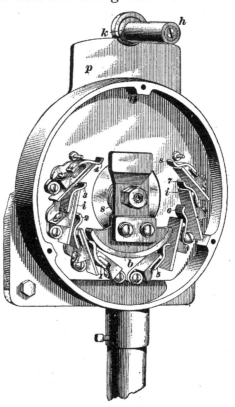

Fig. 41.

When moved to the left, the car ascends, and when moved

to the right, it descends. The arm carries a contact arc b that makes contact with the fingers *1, 2, 3, 4*, etc. when the handle is moved from its central position. The arc b is of such a length that when the handle is moved to its extreme position in either direction, it makes contact with all four fingers on the side to which it is moved. The handle h is held at the central position by means of a spiral spring s, so that if the operator releases the handle, it at once returns to the off position. When the handle is at the off position, the projecting rim k rests in a notch in the plate p, and in order to move the handle, it must first be pulled out against the action of a spring. Insulating pieces i are inserted between the fingers, as shown, in order to avoid short-circuiting.

87. Connections for No. 6 Magnet Controller.—Fig. 42 shows the connections of the No. 6 controller. In this diagram the positions of the switches, resistances, and motor armature and field windings have been arranged so as to make the diagram simpler and easier to follow than if the various parts were located in the same positions that they occupy on the controller. The connections are, however, the same as used on the controller shown in Fig. 40, and corresponding switches are lettered alike. The operation is on the whole very similar to that of the G. S. controller. Terminal x of the operating circuit is connected to the $+$ line and terminal y to the $-$ line. By throwing the small switch c down into the dotted position c'', the car cannot be operated from P'. Switch m is normally open, but it can be thrown so as to connect t and d or t and u. If c is thrown down to the position c'' and m is thrown down so as to connect t and d, magnet A' is energized, and if push button pd is then pressed, current will flow through the coil of C' and the elevator will move down. If m be thrown up so as to connect m and u and button pu pressed, switches B' and C' will be operated and the car will move up. In other words, the small switches and push buttons allow the machine to be operated from the controller while the switch P' is cut off.

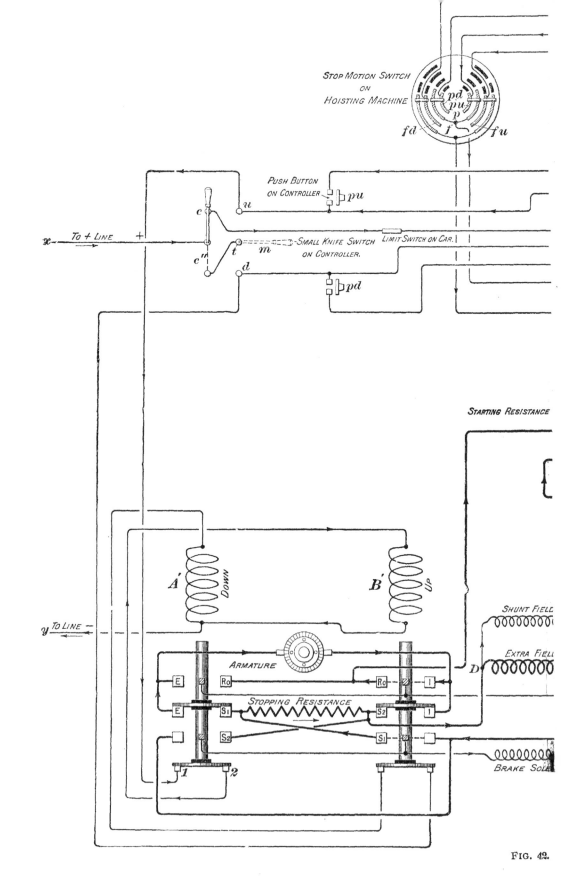

FIG. 42.

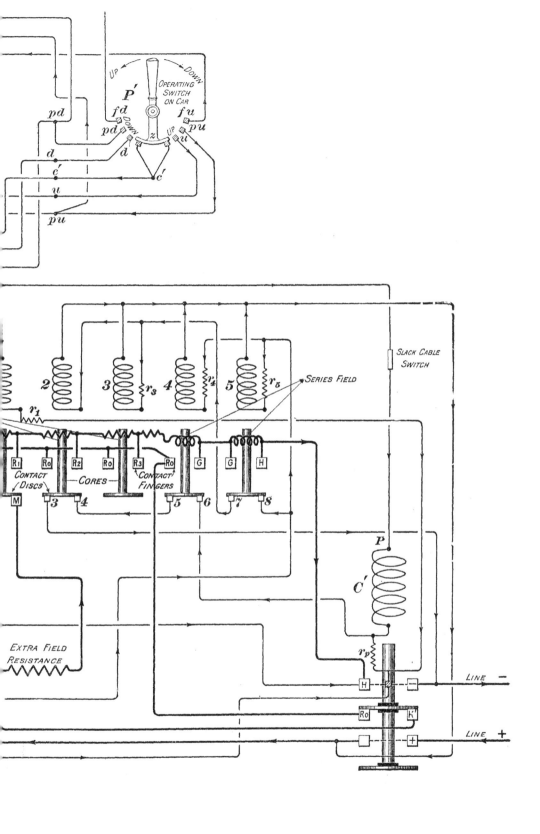

88. Operation of No. 6 Magnet Controller on Starting Position.—Assume that the operating handle is moved to the left so as to bring the arc z in contact with u. Starting from x, the path of the operating current will be as indicated by the arrows through the coil of B' to the negative terminal y. Note that this current passes through the small contacts *1* and *2* of switch A', so that unless switch A' is down, B' cannot be drawn up, and it is impossible, therefore, for both switches to be drawn up together. When z is moved still farther so as to bring it in contact with the finger pu, a current is set up through the operating circuit, which includes the solenoid of switch C'. This current may be traced as follows: x–c–c'–p u–p u to contact pu on stop-motion switch–p–P–through solenoid of C'–*6*–*5*–*4*–*3* to line. This current operates C', which closes the main circuit, releases the brake, and connects the shunt field and extra field across the armature. The various paths of the current are indicated by the arrows, bearing in mind that B' and C' are now up. A powerful magnetic field is provided by the series coils, and as the motor comes up to speed, switches *2*, *3*, *4*, and *5* operate, thus short-circuiting the resistance and the series field. For example, when switch *2* closes, the main current passes from terminal R_o to R_2, thus short-circuiting the first two resistance sections. When the handle of P' is advanced so that z makes contact with finger fu, switch No. *1* is operated. This breaks connection between R_o and M, thus cutting out the extra field. The switches are now all up except A' and the motor runs with a shunt field only; the resistance is all cut out and the motor runs at its maximum speed. When C' is up and when switch *2* operates, contact is broken between the small terminals *3* and *4*, so that the current through C' has to take the path through the resistance r_p to the negative side of the circuit. This resistance is inserted to prevent undue heating of C' and also to save current. Also, when switch *5* is operated, the current through the coils of *2* and *3* is cut off, thus preventing these coils from heating and cutting off the current necessary to energize them. When the core of *2* drops, contact is established again between points *3* and *4*, but in the meantime it has been broken between *5* and *6* so that the current through C' still flows through the resistance r_p. The coils of *1*, *4*, and *5* have considerable resistance in series with them and do not overheat. It is, of course, necessary that these three should remain up while the motor is running, otherwise the extra field resistance, and series fields would not be cut out, while with switches *2* and *3* it is not necessary that they should remain up after sections *2* and *3* of the resistance have been cut out. The voltage at which switches *2*, *3*, *4*, and *5* operate is adjusted by means of the resistances r_3, r_4, r_5, switch *2* having no resistance and, therefore, operating at the lowest voltage. When P' is moved to the right, switch A' is energized and the elevator descends because the direction of the current through the armature is reversed, while that in the fields remains the same as before. The action of the stop-motion switch is the same as in connection with the G. S. controller and needs no special description.

89. Operation of No. 6 Controller on Stopping.— When P' is moved back, contact is first broken with the fu finger. This drops switch *1* and cuts one section of resistance into circuit as well as connecting points R_o and M, thus cutting in the extra field. When contact is broken with finger pu, all the resistance is cut in and the main circuit is opened because switch C' is dropped. In fact, the switches *2*, *3*, *4*, and *5* will likely operate before contact is actually broken between z and pu, if the operating switch is not moved too quickly, because the cutting in of the first section of the resistance and the extra field will lower the speed and thus cut down the E. M. F. applied to coils *2*, *3*, *4*, and *5*. When C' drops, the brake is applied because the circuit through the brake magnet is opened. When the main circuit is opened by C', the armature is still able to send a current around the local circuit E–S_1–through stopping resistance –D–through extra field and extra-field resistance–M–R_o–R_o–R_o–I, because switch B' has not yet dropped.

This generating action through the series field, together with the brake, will soon stop the motor, even if P' is not moved to the vertical position and connection broken with finger u. The car can, therefore, be stopped without the necessity of operating the direction-controlling switch. For example, if the elevator were making an up trip, stopping at each floor, the handle would be moved far enough to break contact with $p\,u$ only, and B' would remain up during the whole trip. When P' is moved to the off position, switch B' drops and the stopping resistance is connected directly across the armature terminals.

AUTOMATIC ELECTRIC ELEVATORS.

90. General Description.—An automatic elevator is one that does not require a regular operator, but is so arranged that it can be controlled by the passenger. These elevators are largely used in private dwellings where the elevator is not used very frequently, and where it would not be desirable or convenient to have an elevator boy.

91. Automatic electric elevators are, with the exception of the controlling devices, similar to other direct-connected electric elevators. There are a number of different styles of them, but the general method of operation is about as follows: A push button is provided at each landing, and in the car there are as many push buttons as there are floors. A passenger at the third floor wishes, say, to go to the first floor. He presses the button at the third floor and the elevator comes up or down, depending on what location it may be in at the time, and when it reaches the third floor it stops automatically, at the same time unlocking the door. The passenger then gets in the car, closes the door of the elevator shaft, and presses the first-floor push button in the car. The car then descends until it reaches the first floor and stops there of its own accord. In the automatic elevator made by the Otis Company, the various devices are so arranged that when the elevator is once started by the passenger it cannot be interfered with by any other person. Also, it is not necessary that the push buttons should remain closed while the elevator is in motion. All that is necessary is to press the button for an instant and then release it. Various safety devices are also introduced; for example, it is impossible to operate the elevator if any of the doors of the shaft are open, and no person on any of the floors can possibly start the elevator if anybody at any of the other floors is getting on or off. We will describe two types of Otis elevator provided with automatic control, and these will serve to illustrate the principle of automatic control in general. The two types are practically the same with the exception of the automatic floor controller, which is geared to the elevator drum.

OTIS AUTOMATIC ELECTRIC ELEVATOR.

92. General Description.—Fig. 43 shows an Otis automatic electric hoisting machine provided with their older type of floor controller. In general appearance it will be noted that the machine is much the same as those previously described in connection with magnet control. About the only difference is the addition of the floor controller shown at C. A spiral contact band a is mounted on an insulating drum, which is moved sidewise by a coarse screw as it revolves, so that the contacts b always press against the band. The various contacts b connect to the push buttons at the different floors. The controller C is used to determine the direction of motion of the car when any given button is pressed, and also to stop the car when it reaches its destination; its action will be understood when the electrical connections are described. The strip b is arranged in spiral form on the drum simply to avoid the use of a drum of large diameter.

93. Magnet Controller for Automatic Elevator.—Fig. 44 shows the magnet controller used with the automatic elevator. A' is the main switch that controls the direction of rotation of the motor. A swinging armature x is hung as shown, between the poles of the two ironclad magnets z, z'. When it hangs in the central position, the circuit is open. When a button is pressed, one or the other of the magnets is excited, depending on the direction in which the car is to move. The armature is drawn over and contact established between the swinging terminals y attached to the arma-

FIG. 43.

ture and the fixed terminals w. The pair of magnets C' cuts out the starting resistance, and magnet B' closes the main circuit and releases the brake. The devices shown at $d, e, f, g, h,$ and k are known as **floor magnets,** and their function is to hold the push-button circuit closed after it has been once pressed, even though the operator releases the push button itself; their action will be explained later. Magnets l, m are provided to prevent interference with the elevator from other floors until the party who is already operating it is through.

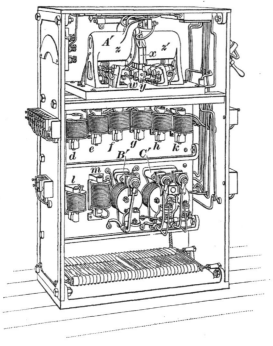

Fig. 44.

94. Connections for Automatic Elevator.—Fig. 45 shows a diagram of connections for an Otis automatic elevator with the style of floor controller shown in Fig. 43. Like the preceding diagrams, the location of some of the parts has been changed in order to make the connections easy to follow. The diagram shows the connections necessary for the control of the elevator from four floors. A' is the reversing switch; the open contacts are mounted on the swinging armature and the shaded contacts are fixed. When coil s is excited, the armature is drawn towards s, and the open contacts make connection with the lower row of fixed contacts, as indicated by the dotted lines. We will assume that with the connections shown, the car moves up when coil s is energized and down when t is energized. Switch B' closes the main circuit; $1''$, $2''$, $3''$, $4''$ are the small floor magnets shown at d, e, f, etc., Fig. 44; 1, 2, 3, 4 are the push buttons on each floor; $1'$, $2'$, $3'$, $4'$ are the push buttons on the elevator; d_1, d_2, d_3, d_4 are door contacts. These contacts are in series and are connected together as shown only when

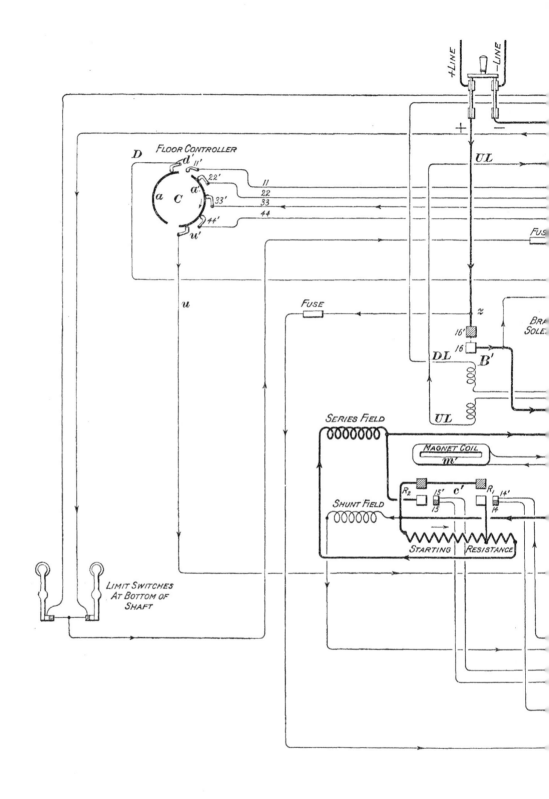

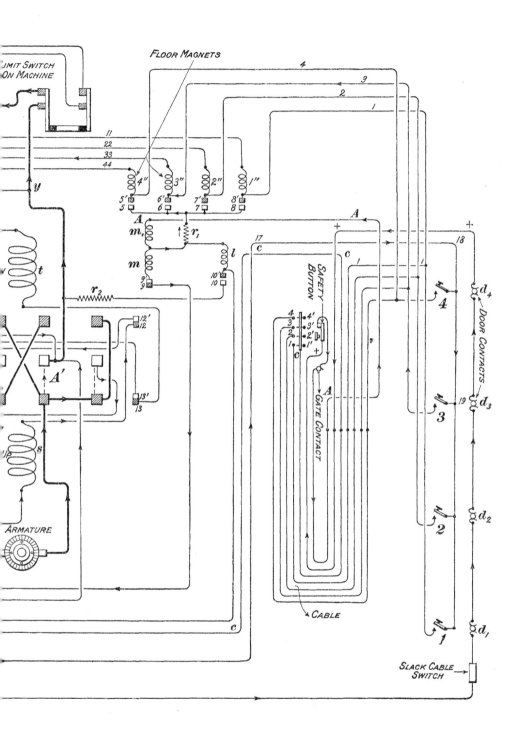

all the doors of the shaft are closed. If any one of these contacts is open, i. e., if any door is not closed, it is impossible to start the elevator. The two switches for cutting out the resistance and the series-field coils are shown at c', and they are operated by magnet m'. C is the floor controller driven from the winding drum. The two segments a, a' represent the strips a shown in Fig. 43. On the drum itself they are put on spirally in order to allow a small diameter of drum, but here they are shown as two arcs, so that the connections may be more easily followed. With the position of the controller shown, the car is at the bottom of the shaft, i. e., at the first floor. As the car ascends, the controller turns as indicated by the full-line arrow, and a' slides from under $22'$, $33'$, $44'$, and u' in succession. At the same time, segment a comes in contact with $11'$, $22'$, $33'$, and $44'$ in succession.

95. When switch A' is pulled by coil t, contact is broken between the auxiliary contacts 12, $12'$, thus opening the circuit through coil s. Also, when A' is operated by coil s, contact is broken between contacts 13, $13'$, thus opening the circuit through coil t. It is, therefore, impossible for both s and t to be excited at the same time. When switch c' makes contact at R_1, contact is broken between 14 and $14'$; when contact is made at R_2, contact is broken between 15 and $15'$. When the armature of magnet $m\,m_1$ is drawn up, contact is broken between points 9, $9'$; when the armature of l is attracted, contact is made between points 10, $10'$. When magnets $1''$, $2''$, $3''$, etc. operate, contact is made between 5 and $5'$, 6 and $6'$, etc. When buttons $1'$, $2'$, $3'$, or $4'$ are pressed, contact is made between wire c and the wire running to the floor magnet corresponding to the button that has been pressed.

96. Operation of Automatic Elevator.—As already stated, the controller C indicates that the elevator is at the first floor. Finger $11'$ is open-circuited, and, pressing either button 1 or $1'$ would not start the elevator, because it is already at the first floor. Suppose a passenger wishes to get on at the third floor and then go down to the first floor. He presses the third-floor button *3*, and the elevator moves up to the third floor and stops there, at the same time automatically unlocking the door of the shaft. After the passenger has entered the elevator and closed the door, he presses button *1'* in the car and the elevator descends to the first floor and stops. The way in which the control is brought about will be understood by following the circuits in Fig. 45.

97. When button *3* is pressed, the operating current, starting from point *z* on the main + wire flows through the slack cable switch, through door contacts d_1–d_2–d_3–d_4, through cable to car and through safety button on car to *A–A–A–m_1–m–9'–9–14'–14–17–18–19*, through push button *3–3*, through floor magnet *3''–33–33*–finger *33'*–strip *a'*–finger *u'–u–u*–magnet *s–12–12'*, through *UL* magnet of *B'*, through limit switch on machine, through limit switches at bottom of shaft, to negative side of the circuit at *y*. This operating current accomplishes several results. In the first place, it closes magnet *3''* so that contacts *6* and *6'* are brought together. This provides a path for the operating current that is independent of the path through push button *3*. The current can now flow along *A A* through coil m_1–r_1–*6–6'* through coil *3''*, and so on as before. Consequently, after button *3* has been pressed the car will start up, even though *3* be released again, and it is not necessary for the passenger to keep the button pressed until the car reaches its destination. All that is necessary is to press *3* long enough to allow *3''* to attract its armature. When the operating current takes the path through *6–6'*, a resistance r_1 is in circuit, as only a small operating current is needed to hold the armatures after they have been attracted. The operating current flows through *s* and one coil of switch *B'*. Hence, the reversing switch is pulled to the up position and the main circuit is closed, thus allowing the main current to flow as indicated by the arrows and starting up the motor. The operating of these switches also allows current to flow through the shunt-field coil and brake solenoid, thus releasing the brake. When coils m, m_1 are energized, points *9*, *9'* are separated, thus breaking all connection between wire AA and wire *17*, *18*, *19*, which leads to one side of all the floor buttons; consequently, as soon as one button, in this case *3*, has been pressed, the buttons on all the other floors are cut out of service, and it is impossible for any other parties to operate the elevator. As the motor speeds up, switch R_1 operates because coil m' is connected across the armature terminals, and this cuts out the greater part of the resistance, at the same time separating points *14*, *14'*. When sufficient speed is attained, switch R_2 operates and cuts out the remainder of the resistance and the series field, at the same time separating points *15*, *15'*.

98. All the time that the car is going up from the first floor to the third, controller C is turning as indicated by the arrow, until, when the third floor is reached, a' slides from under finger *33'*, thus interrupting the operating current and stopping the motor. The motor is stopped by the band brake and no provision is made for a dynamic braking action, as these elevators are not intended for high-speed service.

99. After the car has stopped at the third floor and automatically unlocked the shaft door, the passenger slides back the door, thus opening the operating circuit at contacts d_3 and making it impossible for any persons on the other floors to start up the elevator while he is getting on. After closing the door and thus reestablishing contact at d_3, he presses button *1'*. This allows the operating current to flow as follows, starting from z as before: z–slack cable switch–d_1–d_2–d_3–d_4, through cable to safety button on car–A–A–A–through m_1–l–*15'*–*15*–c–c–c–c–*1'*–*1*–*1*–*1*–*1*–*1''*–*11*– *11*–*11'*–a–d'–D–t–*13*–*13'*–DL, through limit switch on machine, through limit switches at bottom of shaft to y. It must not be forgotten that by the time the elevator has reached the third floor, fingers *11'* and *22'* are resting on contact strip a, and hence are in connection with the wire D that runs to t. Switches B', A', and C' therefore operate as before, except that A' allows the current to flow through the armature in the reverse direction and reverses the motor. Floor magnet $1''$ makes contact between 8 and $8'$, so that the operating current will continue to flow even after the button $1'$ is released. Contacts $9, 9'$ are also separated, so that the push buttons $1, 2, 3, 4$ are cut out while the elevator is in motion. Magnet l is excited and makes contact between 10 and $10'$, thus allowing current to flow to the negative line by the path $A-m_1-l-10'-10-r_2--$, and this current holds 10 and $10'$ in contact, even though the operating current through $1''$ is interrupted by the controller C when the elevator reaches the first floor. When the elevator reaches the first floor, it is automatically stopped by the controller, as already explained, but contacts 9 and $9'$ are still separated and contacts 10 and $10'$ closed, because current still flows through the path $z-d_1-d_2-d_3-d_4-A-A-A-m_1-l-10'-10-r_2--$. The result is that no one can interfere with the elevator because pushes $1, 2, 3, 4$ are cut out. This current through m_1 and l continues to flow until the door is opened, thus breaking the circuit at d_1 and allowing the armature of m and l to drop. After the passenger has gotten out and after the door has been closed again, thus bridging the break at d_1, the elevator may be operated from the other floors, but not before; thus avoiding the possibility of accident while the passenger is getting out.

By tracing out the connections and bearing in mind the action of the controller C, the student will see that the car is under complete control at all times, and that it is practically impossible for any person to interfere with the operation while another person is using it. The unlocking as well as the opening of the doors on these elevators is usually automatic.

OTIS AUTOMATIC ELECTRIC ELEVATOR WITH NO. 2 FLOOR CONTROLLER.

100. General Description.—The style of floor controller shown in Fig. 43 and indicated at C, Fig. 45, has been superseded by a later type shown at C, Fig. 46. Both styles are, however, in use, so that it has been thought advisable to illustrate both of them. The floor controller in Fig. 46 is considerably different in construction, but it accomplishes the same results as the older type; it is mounted on top of the motor and driven by a chain *a* running over a sprocket wheel on the end of the drum shaft.

FIG. 46.

The controller *C* serves also as a limit switch, so that it is not necessary to provide the hoisting machine with the usual traveling nut operating a limit switch. Fig. 47 is a larger view showing one side of the controller. The sprocket wheel *s* is revolved by means of the chain, and by means of reduction gearing turns a shaft *b* through an arc proportional to the total rise of the elevator. On this shaft a number of arms *c* are mounted, each arm carrying a small roller *d*. On shaft *e* a number of cams *f* are loosely pivoted, each cam carrying at its end a cross-contact piece *g*; the shape of these contact pieces is more clearly indicated by the large one shown at *g'*. Each contact has a groove in which the rollers *d* press as shaft *b* revolves, thus forcing up the contact and making connection between the two clips

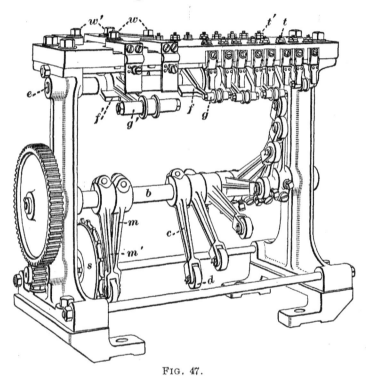

FIG. 47.

with which it is brought in contact. There are two rows of contacts, one on each side of the controller, together with a corresponding number of arms, cams, and cross-contact pieces. The two pairs of large terminals shown at *w, w'* connect to the main circuit, contact being made between them by means of the large cross-contact piece *g'* and a similar one on the other side of the controller. These two main switches are operated by the arms *m, m'*. This controller accomplishes the same result as the one shown at *C* in

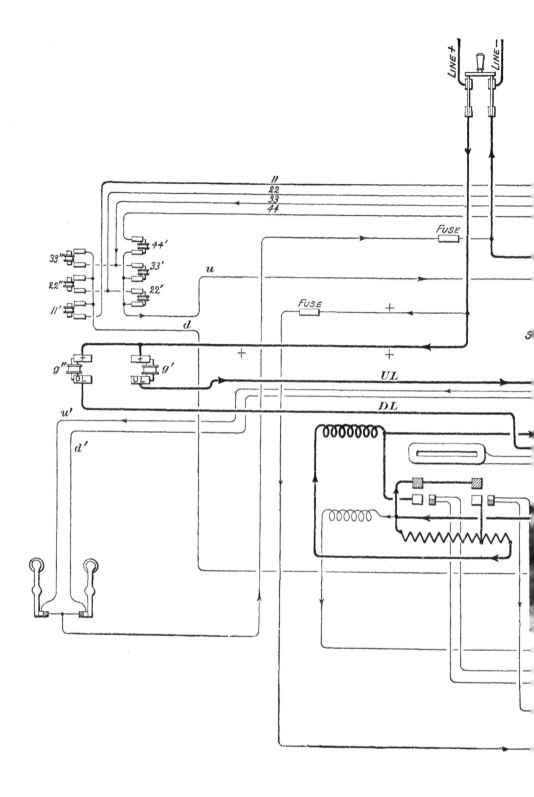

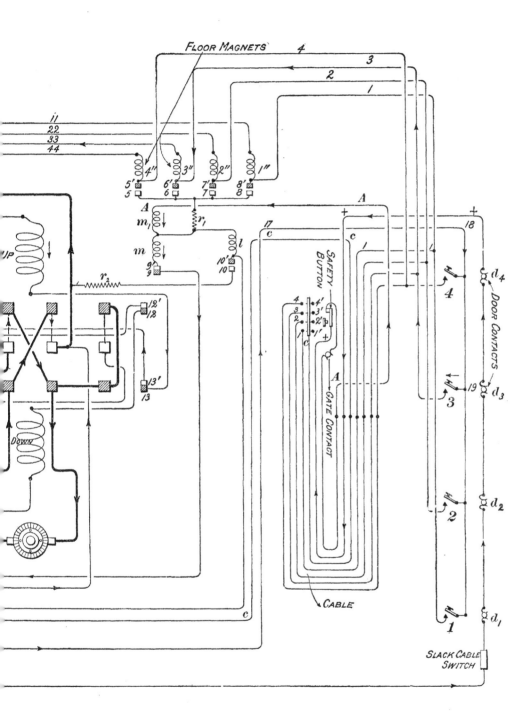

Fig. 43, but uses a series of mechanically operated switches in place of a series of brushes with a sliding contact.

101. Connections for Otis Automatic Elevator With No. 2 Floor Controller.—Fig. 48 is a diagram of connections for this type of controller. It will be noted that it is very similar to Fig. 45, the limit switch on the machine and switch B' being omitted. The up-and-down magnets on switch A' are reversed in position from that in the first diagram, but this is immaterial, as the direction of rotation of the motor may be kept the same in both cases by reversing the armature terminals at the motor. The two large cross-contact pieces on the controller are shown at g', g'', and the small contacts are indicated at *11'*, *33'*, *22''*, etc., there being but three small movable contacts on each side in the diagram, because the elevator is controlled from four floors only. With the diagram as shown, the car is supposed to be at the first floor. All the left-hand contacts of the controller are out and all the right-hand ones are in, connecting the floor magnets to the u line and allowing current to flow through the up magnet of switch A' when any one of the push buttons *2*, *3*, or *4* is pressed. As the car moves away from the first floor, *11'* closes, and when it reaches the second floor, *22'* opens. As it moves away from the second floor, *22''* closes, and when it reaches the third floor, *33'* opens; and so on. When the car reaches its upper limit of travel, switch g' is opened and when it reaches its lower limit, g'' is opened, thus cutting off the main current. When the elevator descends, the switches open and close in the reverse order. The small arrowheads show the path of the operating current when button No. *3* is pressed to bring the car up to the third floor. The large arrowheads show the path of the main current. It will not be necessary to trace these through, since outside of the part through switch C they are practically the same as explained in connection with Fig. 45. Ordinarily, the switch A' would open the main circuit and switches g' and g'' are intended more as a safety device in case A' does not operate.

SPRAGUE-PRATT SCREW ELEVATOR.

102. General Description.—The hoisting mechanism of this elevator differs in a marked degree from those previously described; Fig. 49 shows the general construction. It has no winding drum, the cable being taken up over a number of multiplying sheaves. The hoisting rope H, or rather set of four ropes, passes over the fixed sheaves S' and movable sheaves S and is anchored at A. The motor M revolves a long screw E, which is directly coupled to the motor shaft. On this screw is a nut N, which is not connected in any way with any other part of the mechanism. A crosshead carrying sheaves S is arranged to slide on the base B, and when screw E is revolved by the motor, the nut bears against the crosshead and moves the sheaves S to the right, thus taking up the cable and raising the car. The construction of the nut N and the sheave bearings is such that there is very little friction, and the efficiency of the hoisting mechanism is so high (about 70 per cent. from car to motor) that when the car is descending, the pull against the crosshead revolves the screw and motor in the reverse direction, thus driving the motor as a generator. The sheaves are usually designed to give a multiplication of 8 to 1, so that the amount of rope that the machine takes up is 8 times the travel

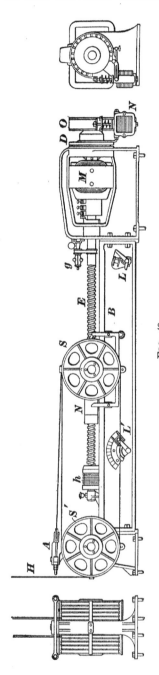

FIG. 49.

of the screw. For high rises and high speed, there is a further multiplication of 2 to 1 on the counterbalance. The ropes lead from the car over the overhead sheaves, down around a sheave on the counterbalance, up to and anchored at the top of the building. The ropes leading to the machine are attached to the bottom of the counterbalance. There are four of these ropes, as indicated in the end view, Fig. 49, two of them passing around the eight sheaves on one side of the machine, and the other two passing around the eight sheaves on the other side. The travel of the car is, therefore, 16 times that of the nut. The screw E is always under tension, no matter what the load on the elevator may be and no matter whether it be moved up or down. This is necessary with this type of elevator because the construction of the nut and screw is such that the pressure between them must always be in the one direction, and the tension on the rope is the only driving power that the elevator has when descending. These machines are not, therefore, overbalanced.

103. Motor.—The motor used with the Sprague-Pratt elevator is of the ordinary direct-current four-pole type with compound field winding. It is mounted at the right-hand end of the machine, as shown at M, Fig. 49, and needs no special description.

104. Transmitting Devices.—The transmitting devices of this elevator are of special interest. The use of the screw, traveling nut, and sheaves makes the action similar in many respects to that of a hydraulic elevator. The sheaves are mounted on roller bearings so as to run with little friction, and the traveling nut is arranged so that the thread of the screw bears against balls. Fig. 50 shows a section of the **ball nut.** Steel balls a are arranged as shown, and when the screw turns, these balls revolve and work their way along through the nut, passing in at one end, traveling through the nut, and returning by way of the channel b in one side. The rolling friction of such a

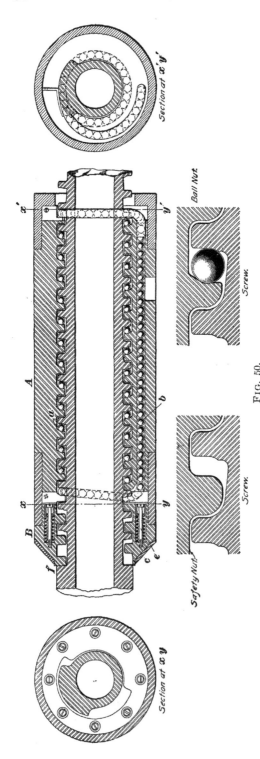

Fig. 50.

nut is very much less than the sliding friction of an ordinary nut. In addition to the ball nut A, there is provided a safety nut B, which is without balls, because under ordinary circumstances there is no pressure taken up between its threads and those of the screw. The safety nut is provided for two purposes, namely: to prevent slack cable and also to hold the crosshead in case the threads on the ball nut or screw should strip. This last contingency is something that never occurs if the elevator receives any kind of inspection, but as these elevators are intended for passenger service, it is advisable to take every possible precaution. When the car is drawn up, there is a thrust between the conical bearing c and the crosshead that carries the movable sheaves, and since the friction of this conical bearing is much greater than the friction of the screw, the nut does not revolve, but travels along the thread, thus pushing the crosshead and raising the car. When the car descends, the pull on the rope runs the screw backwards, and with it the motor, which now runs as a generator. When the pressure on the nut is released, the screw continues to revolve on account of the momentum of the armature, and the cable would be slackened if the nut did not revolve with the screw. As soon, however, as the pressure on bearing c is released, springs e, which are normally compressed, force nuts B and A apart, thus bringing the threads of the safety nut into contact with those of the screw and producing enough friction to make the screw and nut revolve together and thus hold the crosshead stationary. Again, if the threads of the ball nut should wear excessively, or strip, the pressure is taken up on the safety nut, which then revolves with the screw and indicates the defect. A buffer h, Fig. 49, is provided for the nut to strike against when it reaches the limit of its travel corresponding to the lowest position of the car. When it reaches the upper limit, the end of the nut comes up against the shoulder f, Fig. 50, and any further turning of the screw simply causes the nut to revolve with it. The nut shown in Fig. 50 is the later type using a hollow screw of large diameter with $\frac{5}{8}$-inch balls. The balls used on earlier types of the machine were $\frac{1}{2}$ inch in diameter, but this size was found to be rather small. The nut shown contains 320, $\frac{5}{8}$-inch balls. The nut used formerly had 240, $\frac{1}{2}$-inch balls.

105. Thrust Bearing.—The thrust of the screw is taken up by a special form of thrust bearing, which is located at D, Fig. 49, on the back end of the motor frame. The thrust is taken up on a large number of small rolls placed between two hardened steel plates. One plate is carried by the field yoke and the other revolves with the shaft. The small rolls, 180 in number, $\frac{1}{2}$ inch in diameter by $\frac{3}{16}$-inch face, are placed in openings, arranged in spiral form, in a bronze plate. A plate containing these rollers is shown in

Fig. 51; this is placed between the two hardened plates previously mentioned, and the whole thrust bearing is arranged so as to run in oil.

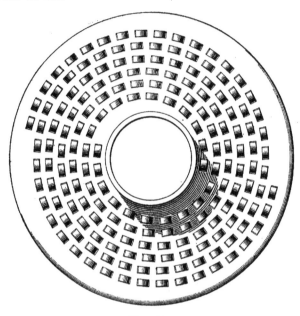

FIG. 51.

106. Brake.—The elevator is provided with a band brake controlled by a solenoid. This brake is shown at O, Fig. 49, and consists of a steel band lagged with wood. The band covers about three-fourths of the circumference of the brake wheel. The solenoid N operates against a spring, so that when the magnet is excited the brake is released, and when it is demagnetized the brake is at once applied by the spring.

107. Limit Switches.—Two limit switches L and L', Fig. 49, are mounted on the base and are operated by projections on the traveling crosshead, so that if the sheaves reach the limit of their travel in either direction, the motor is stopped. Switch L is ordinarily closed, and when the car reaches the upper limit of its travel it is opened, thus opening the main circuit and applying the brake. When the car is descending, switch L' is normally open and when L' is operated at the lower limit it is closed, thus cutting in a resistance across the motor and gradually cutting it out with further motion of the crosshead. A centrifugal governor g is belted to the screw, and if the speed exceeds the allowable limit, this governor opens a circuit and effects an application of the brake.

108. Method of Control.—The method of control used with the Sprague-Pratt elevator is similar in many respects to the magnet-control method previously described. The magnet type of controller might be used with this type of elevator, but many of the Sprague-Pratt machines are equipped with a controller in which resistance is cut out by means of a sliding arm moved by a small *pilot motor*. The closing of the main circuit and the reversing of the motor is accomplished by means of electromagnetic switches very similar to those shown in Fig. 40. The pilot motor is under the control of the car operator and is operated by means of a car-operating switch in a manner similar to that already described in connection with magnet control.

109. Sprague-Pratt Vertical Type Elevator.—Most of the Sprague-Pratt machines have been of the horizontal type shown in Fig. 49. In cases where two or more elevators are required, these horizontal machines are placed one on top of the other, thus economizing space, but a number of machines have been built so that they may be placed vertically in the same way as a hydraulic elevator. Fig. 52 shows the general arrangement of one of these vertical machines. The motor M is at the bottom of the shaft. The fixed sheaves A are mounted just below the lower limit of the counterbalance, and the movable sheaves S travel up and down in guides. The rope running to the sheaves is fastened to the under side of the counterbalance, and there is a multiplication of 2 to 1 between the counterbalance and the car, as indicated. The vertical type has some important advantages over the horizontal type. In the horizontal type, the long screw always tends to sag more or less, thus producing uneven wear. This sagging effect also produces uneven wear on the motor bearings and on the thrust bearing.

FIG. 52.

When the machine is placed in the vertical position, these effects are done away with entirely, and the additional advantage is gained that the weight of the screw, armature, and sheaves tends to counterbalance some of the thrust and thus reduces the effective pressure on the thrust bearings.

FRASER DIFFERENTIAL ELEVATOR.

110. General Description.—This elevator has not as yet been widely used, but as it is very simple in construction and easily controlled, it is probable that it will prove valuable for many kinds of service. It is manufactured by the Otis Elevator Company. The principle on which the elevator operates is an interesting one and will be understood by referring to Fig. 53; A is the car; B and C two pulleys that revolve in opposite directions, as shown. W is the counterweight and D an endless rope passing over the pulleys B, C, and around pulleys E and Y on the car and counterweight. Pulleys B and C are driven by independent sources of power, so that their speed with regard to each other may be changed. If the circumferential speed of B is exactly the same as that of C, it is evident that the rope D will simply pass around over the pulleys and the car will remain stationary. If, however, the circumferential speed of C is made greater than that of B, the rope will be passed over C faster than it is taken up by B and the car will descend. If the circumferential speed of B is greater than that of C, the rope will be taken up by B faster than it is paid out by C, and the car will ascend; the greater the difference in circumferential speed, the greater is the speed of the car. It should be noted that the action depends on the difference of *circumferential speed*, or upon the difference in speed at which the rims of sheaves B and C travel. Pulleys B and C may or may not revolve at the same speed when the car is stationary, depending on whether or not they have the same diameter.

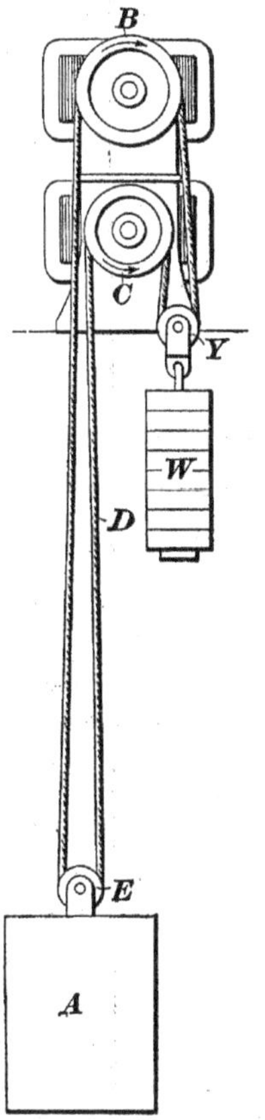

This type of elevator allows the car to be stopped, raised, and lowered without stopping or reversing the driving motors. This, of course, is a great advantage. Usually electric motors are used for driving B and C, though steam engines or a combination of engines and motors could be used. Another advantage of this elevator is that it does not require a winding drum with its accompanying gearing.

111. Fig. 54 shows the general arrangement of an elevator embodying this principle and driven by means of two electric motors. B and C are the two pulleys, the circumferential speed of which is varied by changing the speed of the motors m, m'. The endless rope D is in this case not attached to the car, but runs around a pulley E carried on the bottom of the counterweight W and around the pulley Y carried on the under end of the rope-tightening device N. The ropes L are attached to the car, and after passing over sheave M are attached to the counterweight. The ropes F also attach to the top of the counterweight, and after passing over sheave G and through cross-bar 3, are fastened to cross-bar 4 of the tightening device. By drawing bar 4 down on the threaded rods, the ropes can be tightened to any desired degree. The speed of the motors is controlled from the car, and in making a trip they are not

FIG. 53.

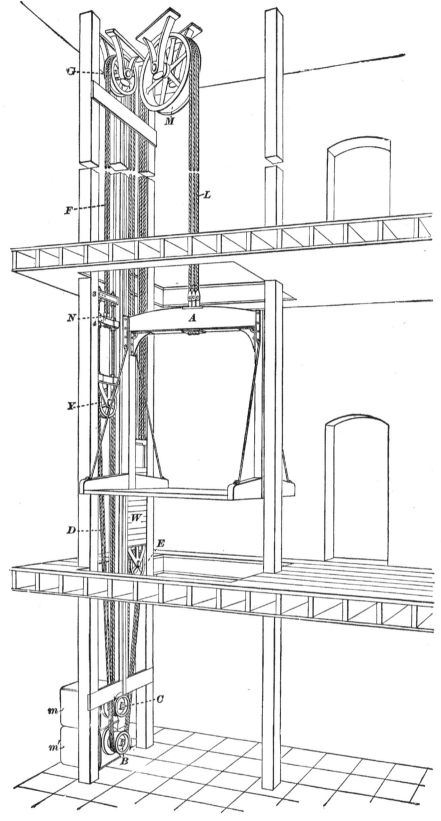

Fig. 54.

stopped when the elevator stops at the various floors; they are merely made to run at the same speed by means of the car controller. While the car is ascending, pulley B runs faster than C, and in order to make it descend, all that is necessary is to make C run faster than B. The variations in speed are readily accomplished by varying the field strength of the motors. The controller is arranged so that when the handle occupies the central position, the speed of both motors is alike and the car is stationary, and when moved to either side of the center, the speed of either one or other of the motors is changed. A small auxiliary-operating handle is also provided in connection with the main handle, so that by pulling up on it the operator can stop both motors when the elevator is not in use.

ELEVATORS.

(PART 3.)

HYDRAULIC ELEVATORS.

INTRODUCTION.

1. Hydraulic elevators are still considered by the majority of engineers as being the most suitable for large passenger-service plants with their high lifts and great speeds, although the electric elevator has since its advent become a powerful competitor. The hydraulic elevator is intrinsically safe, reliable, smooth-acting, and under perfect control. It requires comparatively less care in operation than the electric elevator, the mechanism being very simple. The cost of maintenance is small, the wearing parts being few and easily and cheaply replaced.

On the other hand, the hydraulic elevator is cumbersome, requiring much space, especially—and this is the case in most large plants—where the water pressure available is not high enough for direct use in the elevator cylinders, so that the installation of steam pumps, reservoirs, or tanks, and the necessary piping becomes necessary, not mentioning a boiler plant, which we may assume, for the sake of the comparison, as being already in existence for other than the elevator service. Thus, the first cost is great compared with that of an electric elevator plant, which, in case the right current is already available, either from a central station or from an isolated lighting plant in the building, consists of

§ 39

For notice of copyright, see page immediately following the title page.

the elevator machine only, which may even be placed on top of the hoistway, and in case the current must be generated expressly for the elevator, of an additional steam-engine- or gas-engine-driven dynamo, but no cumbersome tanks or piping. The installation is thus simple and cheap, the space needed but small. There are advantages, then, in both systems, and which one to select depends on many circumstances which must be weighed against one another by the architect and owner, but not by the operating engineer, who should have no prejudice against the one or the other, but should be equally familiar with both.

PLUNGER ELEVATORS.

SERVICE.

2. The simplest kind of hydraulic elevator is the **direct-acting** or **plunger elevator**. It is also the oldest kind of hydraulic elevator and has been used for a long time, both for freight and passenger service, for short lifts. It has been until recently considered unsuitable for high lifts and high speeds, and is therefore found installed in great numbers as yet only for sidewalk lifts, slow freight elevators, and similar service. Lately, however, the possibility of using this type of elevators for greater lifts and speeds has been recognized, and it is safe to predict that they will be more frequently installed than before and for even the severest service.

CONSTRUCTION.

3. Fig. 1 shows a plunger elevator made by Morse, Williams & Co., of Philadelphia, Pennsylvania, for short lifts.

4. Motor.—The motor in this machine consists of a vertical cylinder A sunk into the ground below the bottom of the hoistway and a plunger P. The cylinder is closed at

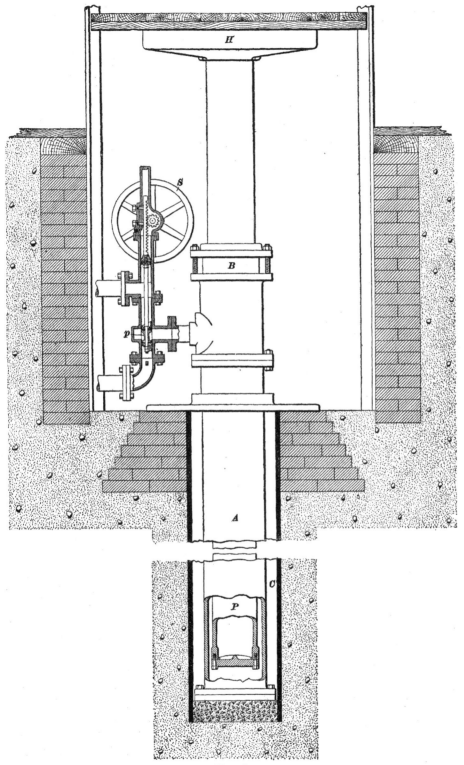

FIG. 1.

the bottom and has an enlarged head above ground containing the stuffingbox B and an opening for the pipe through which the water under pressure enters the cylinder and is discharged from it. The cylinder must, of course, be sunk plumb. If the subsoil is soft earth, it is first necessary to sink a steel pipe C, called the **casing,** through which the earth is removed. When the subsoil is rock no casing is required, the hole for the cylinder being drilled.

For high lifts the cylinder is made up of sections. The Plunger Elevator Company, of Worcester, Massachusetts, use steel tubing, which they square up and thread in the lathe, connecting the sections by means of couplings. This insures a perfectly straight cylinder. Before burying in the ground the cylinder is tested and given a coat of preservative paint.

The plunger when required to be long is also made up of sections of steel tubing. Fig. 2 shows the special joint used by the company named above. The plunger is turned to uniform size and polished.

FIG. 2.

5. Transmitting Devices.—As the car rests directly on top of the plunger, there are no transmitting devices, such as drums, ropes, and sheaves. The car is fastened to the plunger, which is provided with a head H for the purpose. The head shown in Fig. 1 is simply a cast-iron plate clamped to the plunger. This arrangement, while sufficient for unbalanced small elevators, would be dangerous for large counterbalanced ones, inasmuch as should the connection between the head and the plunger give way, the counterweights would jerk the car upwards against the overhead work. Great care is, therefore, taken in balanced elevators of this kind to make the aforesaid connection very rigid and reliable. The manner in which this is done by the Plunger Elevator Company is shown in Fig. 3.

The plunger has a flange formed on its upper end that fits into a corresponding recess of the head *H*. The latter, in turn, is securely bolted to the framework of the car platform. Besides this flange connection a second security

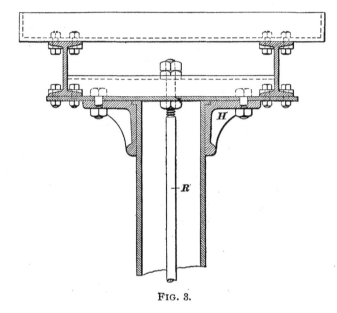

FIG. 3.

against the parting of the car and plunger is provided by a tie-rod *R* which runs all the way through the plunger, through the bottom of the same, and through the framework of the car platform. Instead of the rod *R* a loop of galvanized iron rope is often used for the same purpose.

6. Counterbalancing.—Low-lift plunger elevators are generally not counterbalanced at all. High-lift elevators are counterbalanced, but not overbalanced, since the power acts only on the up stroke of the plunger. Enough of the weight of the car and plunger is left unbalanced to secure the descent of the car at the proper speed when empty. The upward pressure of the water on the plunger gradually diminishes as the plunger goes up by an amount corresponding to the increasing height of the water that displaces the plunger. To equalize this change of pressure, the counterweights are suspended from cables of such size that the weight per each foot of their length passing over the overhead sheaves will be equal to half the weight of 1 foot in height of water displacing the plunger.

7. Controlling Devices.—The controlling devices consist simply of a balanced three-way water valve operated by a simple shipper rope, or a shipper rope in connection with some more elaborate operating device. The simple shipper rope is generally used with the smaller machines, while an operating device of more elaborate form is used for the larger machines.

The valve in a hydraulic elevator constitutes the only controlling device, being power control and brake at the same time. As a power control it shuts off the power at the will of the operator; as a brake it is so designed as to shut off the water gradually by throttling. This object is most easily attained by a piston valve, which type of valve is used exclusively. Thus, while there is no brake in the common meaning of the word in hydraulic elevators, it is, nevertheless, there as in any other elevator, but in a different form.

This identity of power control and brake is one of the intrinsically valuable features of the hydraulic elevator, since by opening the water passages more or less, the speed of the car can be regulated to a nicety and in harmony with the load it carries, which feature is not easily attained, if attained at all, in any other kind of elevator. Generally the valve is proportioned by the installators so that when fully open it will give the empty car the maximum speed permissible; but by the use of stops the valve throw can be adjusted to any car speed. Such stops are generally in the shape of knobs or buttons clamped to

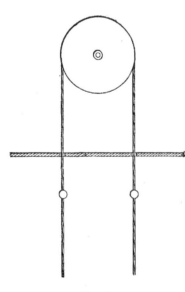

FIG. 4.

the shipper rope and striking against some fixed projection, as shown in Fig. 4. These stops are called **back-stop buttons.**

8. The valve used on the Morse, Williams & Co.'s elevator is shown in section in Fig. 1. A peculiar feature of this valve is the shape of the piston p, which is seen to be wedge-shaped, in consequence of which the water passes to and from the machine gradually and without shock. The operation of the valve will be readily understood from the drawing; on shifting it one way by means of the shipper rope passing over the sheave S, water flows from the *supply* into the *machine* and exerts a pressure on the plunger, lifting it and the car. By shifting the valve in the opposite direction, communication is established between the *machine* and the *discharge*, and the elevator descends. In the intermediate position, the valve shuts off all communication of the machine with the supply and discharge and the elevator is at rest, the plunger being supported on a column of water confined in the cylinder.

9. In larger machines the controlling valve is preferably moved by a motor piston, which is operated by a **pilot valve.** The pilot valve is in turn controlled by the shipper rope from the car. The arrangements of pilot valves and main valves differ in different installations, but are easily understood in every case by inspection. We shall encounter the pilot valve again in connection with piston elevators, when descriptions and drawings of several types will be given and their purpose explained.

10. Safety Devices.—The plunger elevator is one of the safest elevators. The ordinary knobs or buttons used on the shipper rope as limit stops are the usual motor safeties provided, and even should the limit stop fail to operate the valve at the top of the run, the counterweight would reach the ground and the car stop ; should the limit stop fail to operate at the bottom of the run, the car would simply come to rest on the cylinder. In order to avoid damaging the cylinder head in case this should happen, buffer springs are

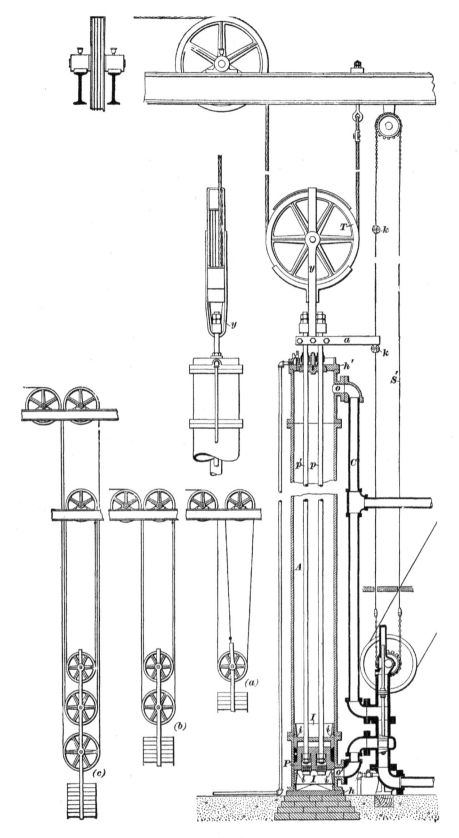

FIG. 5.

often placed on top of the cylinder head, especially when the speed of the elevator is considerable. Car safeties, which are essential on all other elevators, are not needed in plunger elevators, for the car cannot fall, since the plunger always rests on a column of water that is driven out through comparatively small openings; it may, however, in case the valve should fail to operate, attain a speed that would be undesirably great, though not dangerous. To provide against this, the simple expedient of putting in the discharge pipe a throttle valve controlled by the pressure corresponding to the velocity of the exhaust is resorted to. The Otis Elevator Company also place automatic step valves in the supply and discharge pipes, which are closed by the car at the limits of its travel, and hence are independent of the operating device or the controlling valve.

PISTON ELEVATORS.

ADVANTAGES.

11. While the plunger elevator treated in the previous articles is simplicity itself, it has some disadvantages. The hydraulic cylinder and plunger must have a length equal to the lift, and for each trip of the car a volume of water is used equal to the area of the plunger multiplied by the lift. In the piston elevator, by introducing multiplying sheaves the hydraulic cylinder can be made considerably shorter, the volume of water used is, however, about the same for the same work, as the diameter must be increased proportionally. There are two types of piston elevators; in one the cylinder is *vertical* and in the other *horizontal*.

VERTICAL HYDRAULIC PISTON ELEVATORS.

12. The vertical type is considered better than the horizontal type, and is always installed when circumstances will permit, chiefly for the reason that generally headroom is more available than floor space. Fig. 5 is a section through the cylinder, piston, and valve of a simple machine of this kind, as built by Morse, Williams & Co., of Philadelphia, Pennsylvania.

13. Motor.—Following up the various parts, we have as the motor a cylinder A and piston P, the former consisting of a number of cast-iron flanged sections bored and faced true and bolted together at their flanges; a bottom head h and a top head h', which latter contains the stuffingboxes for the piston rods p and p'. The cylinder has two openings o and o', at the top and bottom, respectively.

14. Transmitting Devices.—The transmitting devices consist of wire ropes running over sheaves, one or more of which are carried in a yoke y attached to the piston rods, while the others are supported in bearings on overhead beams. The main figure shows but one **traveling sheave** T; the car in this case moves twice as fast as the piston, and the elevator is said to be geared in the **ratio** 2 : 1. Fig. 5 (*a*), (*b*), and (*c*) shows the arrangement of sheaves for the ratios 3 : 1, 4 : 1, and 6 : 1, respectively.

15. Counterbalancing.—As we shall see presently, matters are arranged in most vertical elevators so that the cylinder is always full of water. This gives rise to an advantage in counterbalancing. The piston is always carried on a solid column of water and thus forms a counterweight that will come to rest at the moment when the power is cut off, that is, the flow of water stopped; contrary to a free counterweight, it will thus not produce a tendency to teeter the car up and down by its momentum when the power is suddenly cut off. The counterweights in these elevators are, therefore, preferably placed wholly or at least partly on the piston or piston rods, as shown in Fig. 5 (*a*), (*b*), and (*c*). As the power acts only on one side of the piston, the counterweights must be less than the car weight by an amount sufficient to make the car descend at the **proper speed when empty**.

16. Controlling Device.—The controlling device consists of a balanced three-way valve operated by a shipper rope in the usual manner, the rope being provided with back-stop buttons. The action of the motor under its control is as follows: The space of the cylinder above the piston is always filled with water under pressure, the supply pipe being connected with this space directly through the **circulating pipe** C. The other end of the circulating pipe is connected with the space of the valve chamber between the two valve pistons. If the valve pistons be moved downwards, so as to bring the upper valve chamber and thus the space of the cylinder above the piston into communication with the space below the piston, there will be the same water pressure on both sides of the piston. The car, being heavier than the piston with the counterweights, will cause the latter to ascend while it is itself descending, and will force the water from above the piston through the circulating pipe into the space under the piston. For the ascent of the car the valve pistons are raised so as to put the space of the cylinder below the piston into communication with the discharge pipe; there is then pressure only on top of the piston, and the same descends, raising up the car. In the position shown in Fig. 5, the valve closes the space below the piston against both the supply and the discharge, so that the piston is held between the water pressure from above and a confined water column from below.

The object of making the water circulate from the top to the bottom of the piston is primarily to make the effective pressure on the piston the same at all points of the stroke, which otherwise would not be the case. Imagine that the cylinder was open at the top and bottom and the piston at the top of its travel, and that water be poured on to the piston from above; then the latter would descend under the influence of the weight of the column of water above the piston, which would be nothing at first, but would gradually increase towards a weight equivalent to the total contents of the cylinder. Imagine, now, that the space below the piston is filled with water, the piston again being at the top; then the column of water underneath it will exert a suction on the piston corresponding to the height of that column, as long as the column is not higher than 34 feet, which suction will gradually decrease to nothing as the piston descends. Thus by having the space below the piston filled with water the same net force is exerted on the piston at all points; for, while the pressure of the water above it increases, the suction of the water below it decreases at the same rate. In concise technical terms, then, the object of the circulation of the water from the top to the bottom of the piston is to balance the head of the water above the piston.

17. Safety Devices.—The safety devices consist of the usual *car safeties* used for suspended cars and *motor safeties*. Limit stops take the shape of knobs or buttons k, k, Fig. 5, on an endless rope S', which are operated by a projecting arm a on the piston rod. The top and bottom heads would of course stop the travel of the piston either way, but it would not be safe to intrust them with that duty, as breakage may result by the piston striking them. The latter should not, therefore, ordinarily travel so far as to strike the heads. There being a possibility, however, that this might occur through a failure of the valve to operate, the piston is provided with an **apron** I on each side; each apron has a number of holes i, i through it and partially closes the ports o or o' and thus reduces the speed of the piston before it reaches the heads. The holes i, i allow the water to enter on the return stroke.

18. It can easily be understood that every elevator should be started and stopped gradually to avoid shocks, and that there always exists the danger of overthrowing the controlling device beyond the neutral point.

Referring to Fig. 5, it will be understood that when the piston is going down the car is ascending, and if the valve is suddenly closed, the flow of the water from the space below the piston through the discharge pipe is suddenly stopped. The momentum of the piston and car will tend, however, to continue the motion, resulting in a thud of the piston against the column of water thus confined. To avoid this **water ram,** as it is called, it is good practice to interpose in the discharge pipe between the cylinder and valve a **relief valve** r, as shown in Fig. 6, which is a drawing of an **Otis vertical elevator** of much the same design—with the exception of some details, to which we shall refer below—as that shown in Fig. 5. The danger of producing a shock by the careless handling of the operating device on the down trip of the car is not so great, inasmuch as the column of water above the piston is not confined in the cylinder on closing the valve, being always in communication with the supply pipe and through it with the pressure tank and its air cushion. A relief valve for the down trip is, therefore, deemed superfluous.

19. Pilot Valves.—For high-speed hydraulic elevators (600 feet per minute and more), the insertion of the relief valve is not sufficient to guard against shocks, it being extremely difficult to start and stop gradually by operating the main valves directly; nor is it possible to regulate the speed readily by opening the valve more or less, so that one of the most valuable features of the hydraulic elevator is curtailed. This has led to the introduction of the **auxiliary,** or **pilot, valve,** already referred to in Art. **9.** Such a valve as built by the Otis Elevator Company is shown in Fig. 7, of which the following is a brief description:

Contrary to the direct-operated valves shown in Figs. 5 and 6, the main valve V, Fig. 7, composed of the pistons v and v', is not balanced, but the upper piston v has a larger area than the lower double one v'; the valve is, therefore, also called a **differential valve,** there being always a pressure against the under side of the upper piston v depending on the difference between the areas of the pistons v and v'. On a bracket B fixed to the main valve casing is supported the auxiliary, or pilot, valve W, which is simply a piston valve of small dimensions; the casing of this valve has an inlet w connected with the circulating, or supply, pipe and

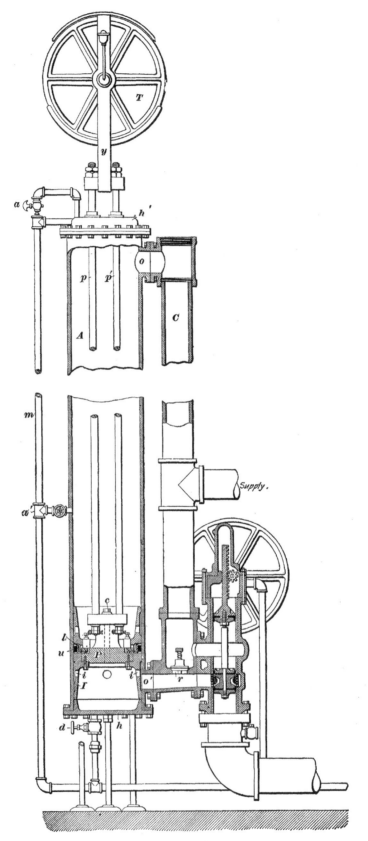

FIG. 6.

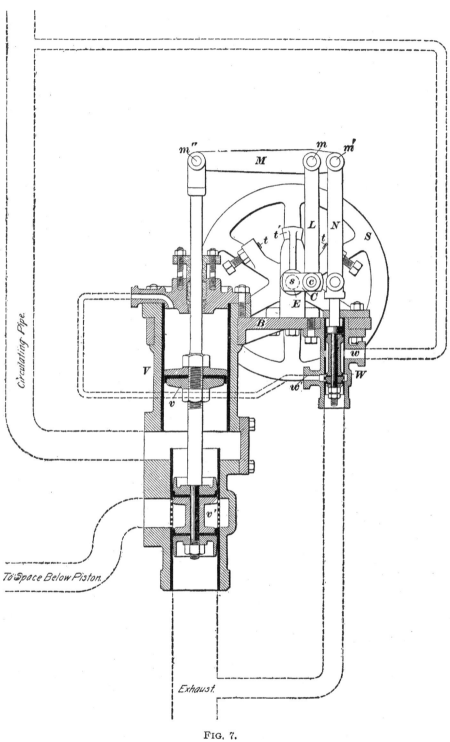

Fig. 7.

an outlet w' connected by a pipe to the space above the upper piston of the main valve, as shown. In the position of the two valves shown in the illustration, the communication between them is shut off, the pilot-valve piston covering the outlet port. The upper space of the main valve is filled with water wholly confined, so that the tendency for upward motion of the piston v is checked in a position where the lower piston cuts off the circulation of the water, when, as we know, the elevator is at rest. By lowering the pilot valve, communication is established between the supply, or circulating, pipe and the space of the main valve above the piston v, which presents its whole area to the incoming water; as the upward pressure below it is less, owing to the difference between its area and the area of the lower piston v', it will descend with the effect of allowing a circulation of water from the top to the bottom of the cylinder, so that the car descends. If, now, the pilot valve be brought back into the position shown in Fig. 7 (the main valve being in its lowest position for the down trip of the car), it would check any farther downward motion of the main valve and the same would thus remain set for the down trip. Again, if the pilot valve were raised beyond the position shown in Fig. 7 (the main valve still being in its lowest position), the space above the piston v would be connected with the exhaust and the main valve would ascend and keep on ascending to the neutral position (elevator at rest) and beyond it (elevator descending), unless the pilot valve be brought back to the neutral point.

Thus, if no provision be made further than described, it would be necessary, in order to stop the car during a downward trip, to throw the pilot-valve operating device completely over, to wait until the elevator came to a stop, and then to throw the device into the central (neutral) position. The same complicated operation would be required for the upward trip.

20. To avoid the complicated operation mentioned in Art. **19,** the two valves are so connected by a system of linkwork that the pilot valve closes automatically without affecting the operating device in the car (shipper rope, lever, or hand wheel) when the main valve reaches its extreme upper or lower positions. This is brought about in the following manner: The shipper sheave S is mounted on the bracket B, its shaft s carrying a crank C, the crankpin c of which is connected to a double-armed lever M by a link L and a pin m. To the right of the pin m is another pin m' that serves as a pivot for a link N, which is connected at the other end to the stem of the auxiliary valve. A third pin m'', to the left of the pin m, connects the lever M with the main-valve stem. Stops t, t, t' on the shipper sheave and its stationary bearing, respectively, limit the motion of the crank C.

The operation is as follows: Starting, as before, from the position of the valves shown in Fig. 7, the piston v is held stationary between the water pressure from below and the confined water above, so that the pin m'' forms the pivot of the lever M when we move the shipper sheave to the right. The crank C then pulls down the lever and with it the pin m', link N, and the pilot-valve stem, thus lowering the pilot valve to the position in which it admits water into the main valve, which then moves downwards. As soon, however, as it commences to move, it raises up the pilot valve, the crankpin c, as well as the link L and the pin m being now stationary, which latter then serves as the pivot for the lever M. The leverage is so proportioned that by the time the main valve has reached its lowest position the pilot valve will be closed, that is, it will have returned to the position shown in Fig. 7, checking further motion of the main valve, the crank C, however, remaining in its lowest position. If it is now desired to stop the car during its down trip, the sheave, and with it the crank, is brought back to the neutral position. The pin m'' being, now, once more the pivot for the lever M, the pilot valve is raised above its neutral position, the main valve rises, and by the time it has risen far enough to shut off circulation of the water it has dragged the pilot valve back to its neutral position. All parts are now again placed as shown in Fig. 7 and the cycle may be repeated.

21. Though it takes many words to describe it, the operation of this valve is very simple and reliable. The operator may with impunity throw the operating device quickly from its neutral position to the right or the left, that is, for "up" or "down," without affecting the gradual, measured motion of the main valve, which is the purpose of the pilot valve. Moreover, it will be understood that the pilot valve allows a perfect regulation of the speed of the car. For by throwing the operating wheel or lever on the car over only part of its full swing, the pilot valve will make only part of its travel and, consequently, will be brought back to its neutral position by the action of the main valve before the latter has completed its full stroke, thus leaving the main valve but partly open, whereby the flow of the water is throttled.

22. Independent Top and Bottom Stop-Valve.—In connection with a pilot valve, the ordinary kind of limit stop shown in Fig. 5 operating the valve directly cannot be used, for the piston or car will still be moving, while the quick move of the pilot valve has long been completed. It becomes necessary, then, to introduce an independent valve for stopping the car at its limits of travel. Such a valve is shown in Fig. 8 at Q, and its construction and operation are as follows: Into the passage leading from the space below the elevator piston to the exhaust, a cylindrical shell q having three passages is inserted, of which the upper passage leads to the relief valve (see Art. **18**). Either of the two passages t and t' may be closed by the rotary valve, shown to an enlarged scale in Fig. 9, which consists of a spindle s passing through stuffingboxes of the valve casing and carrying a valve body v composed of a sleeve and flanges fitting the inside of the shell q. The flanges of the valve body are notched out to receive the valve proper w, which fits with considerable play in the notches, as shown

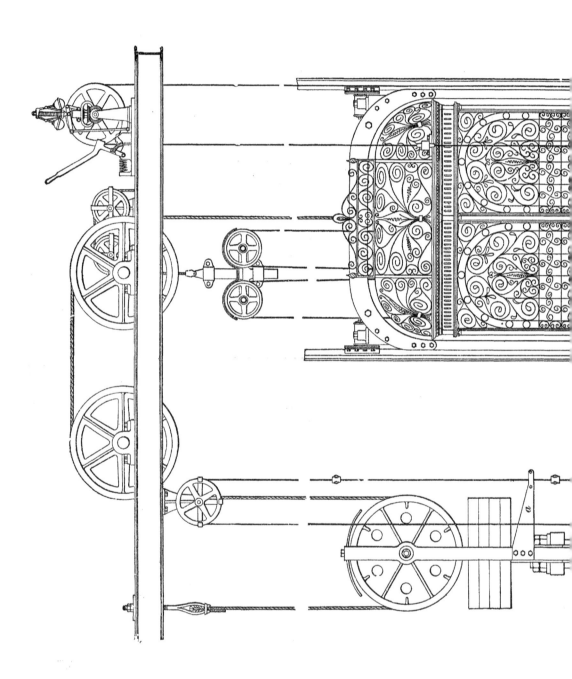

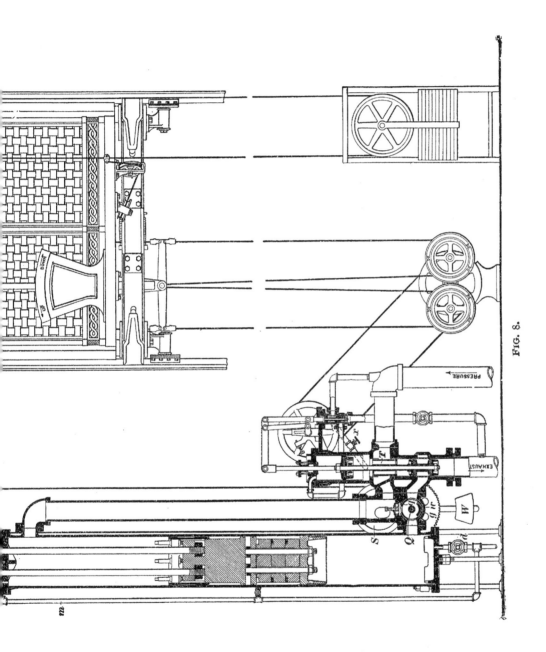

Fig. 8.

in Fig. 9. The valve spindle carries on the outside of the casing a gear-wheel g, Fig. 8, actuated by a weight W that tends to keep the valve in the neutral position shown in Fig. 8. The gear g meshes with a smaller gear attached

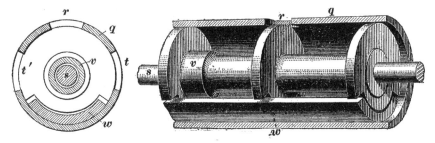

FIG. 9.

to the shaft of a rope sheave S, which is actuated by an endless rope passing over an idler above and carrying the usual stop buttons. Now, when the piston nears its lowest position, the arm a on the piston rod strikes the lower stop button; the sheave S swings right-handed and the valve w turns left-handed, covering up the right-hand port or passage t, thus shutting off gradually the communication of the cylinder with the exhaust and stopping any farther downward motion of the piston, even if the operator has neglected to move the pilot valve.

In order that the elevator may start again upon reversing the pilot valve, the valve w of the rotary stop-valve has a certain play, as already stated. Thus, imagine the piston in its lowest position and the valve w covering up the port t; it will then be pressed against its seat (the shell q) as long as the main valve uncovers the exhaust or is in the middle position, by reason of the pressure above the piston. But as soon as the main valve is reversed, so as to open communication between the spaces above and below the piston through the circulating pipe, there will be an excess of pressure against the outside of the valve w of the rotary valve, due to the unbalanced weight of the car. The valve w will then be lifted off its seat and will allow water to pass below the piston, which then commences to rise. Presently, the arm a will leave the lower stop button and the rotary valve will swing back to its neutral position by virtue of the weight *W*, Fig. 8. Similar action takes place at the extreme upper position of the piston.

23. Throttle.—A difference will be noticed in the construction of the main valve between that shown in Fig. 7 and that shown in Fig. 8, there being interposed in the latter case a metal sleeve between the upper single piston and the lower double one. This sleeve, which is called the **throttle** and is designated by *T*, Fig. 8, is fastened to the valve rod, or stem, and in its neutral position shuts off the supply from the space between the valve pistons. Otherwise, the connections and passages are the same, the supply being in constant communication—except when shut off by the throttle—with the circulating pipe by two branch passages leading from the annular chamber around the throttle to the circulating pipe. In order to show this clearly, a horizontal section through the middle of the throttle and its casing is given to an enlarged scale in Fig. 10 (*a*).

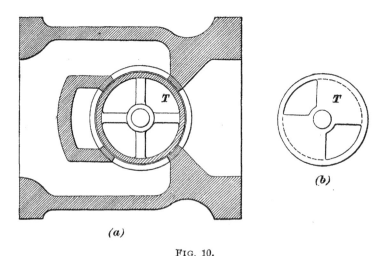

Fig. 10.

The purpose of the throttle is a threefold one. (1) It serves, if carefully adjusted, to deaden the noise occasioned by the circulating water. (2) It serves as a brake while descending, in case of an extra load on the car, preventing it from attaining undue speed. This is accomplished by the top of the throttle sleeve being partly closed, as shown by the plan view given in Fig 10 (*b*), thus allowing only a small amount of water to pass through, that is, throttling it. (3) If any pipe or connection between the supply and valve should break, the water cannot back up from the circulating pipe out through the supply port faster than it can leak around the outside of the throttle.

24. The throttle is but loosely fitted to its seat, or lining, so that there is always some leakage around it, otherwise the elevator could not be started from its position of rest, since there would be no outlet for the water between the large and small piston of the differential valve while descending, and to the inlet while ascending. This leakage is sometimes solely depended on to give the differential valve the initial start, but oftener a by-pass pipe x, Fig. 8, leads from the supply chamber of the pilot valve to the space under the upper piston of the differential or main valve. This by-pass pipe is provided with a globe valve, by means of which the rapidity of the initial start can be regulated.

25. Double-Power Vertical Hydraulic Elevator.—In modern office buildings most of the time a light load is carried, but at times a much heavier load must be raised. A double-power elevator is then often installed, using two pressures in its normal operation, a low pressure being used for a light load and a higher pressure for a heavy load. The higher pressure is obtained from an extra high-pressure tank, and a special valve is used that permits the elevator to be run at will either with the ordinary low pressure or with the high pressure. Such a valve, built by the Otis Elevator Company, is shown in Fig. 11. The upper valve v is, in this case, a piston valve straddling in its neutral and lowest position the high-pressure port. The throttle T has ports t, t. When the valve stem is moved down, the water circulates as in the ordinary hydraulic machine and the car descends. When the valve stem is moved up, the discharge is opened, as in the ordinary machine, and as the low-pressure inlet is connected to the top of the cylinder through the ports t, t and the circulating pipe, it causes the car to ascend. For heavy loads the valve stem is raised still farther until T comes opposite the high-pressure inlet r, and this opens the passage for the high-pressure water into the top of the cylinder through the ports t, t and the circulating pipe, thus giving the car the full benefit of the high pressure. Since in this position of the valves both the low- and high-pressure supply are connected with the circulating pipe and thus with each other, the water would flow from the high-pressure tank to the low-pressure tank, were it not for a check-valve C inserted in the low-pressure supply pipe, as shown.

26. Non-Circulating Systems.—In the vertical-piston elevators thus far described, the distinguishing feature was the circulation of the water from above the piston to the space below it during the descent of the car. As we have seen in Art. **16**, the principal object of this arrangement was the balancing of the head of the water above the piston. Incidentally, certain advantages in counterbalancing were obtained by it also (see Art. **15**).

It is considered in practice that a ratio of 8 : 1 is the limit for vertical machines. In certain designs, however, the ratio has been carried much above this value for the purpose of making the cylinder very small. Now, since the head of the water becomes less and less the shorter the cylinder is made, it becomes unnecessary to balance it when the ratio is very high, say 10 : 1, for instance. In such cases the circulating pipe is dispensed with; the water then enters and leaves on one side of the piston only and one end of the cylinder is left open to the atmosphere. Fig. 12 shows an elevator of this kind made by The Whittier Machine Company, of Boston. There are quite a number of these machines in operation. The ratio of the particular machine illustrated is 10 : 1, there being a set of five fixed and five traveling sheaves on each side of the cylinder;

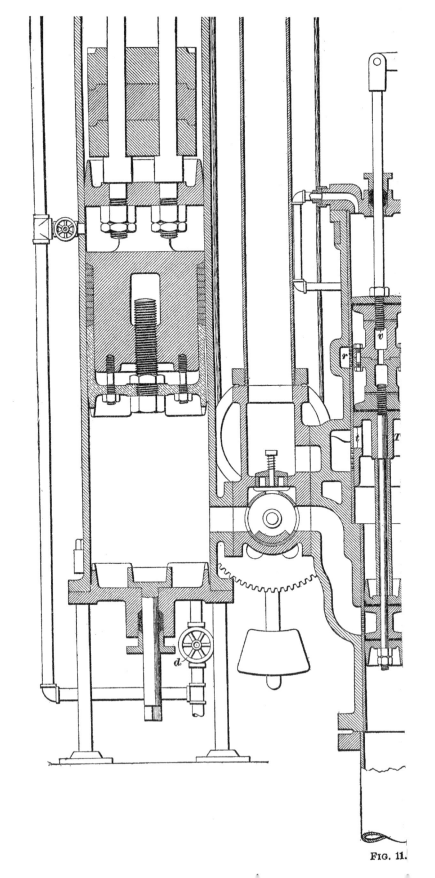

FIG. 11.

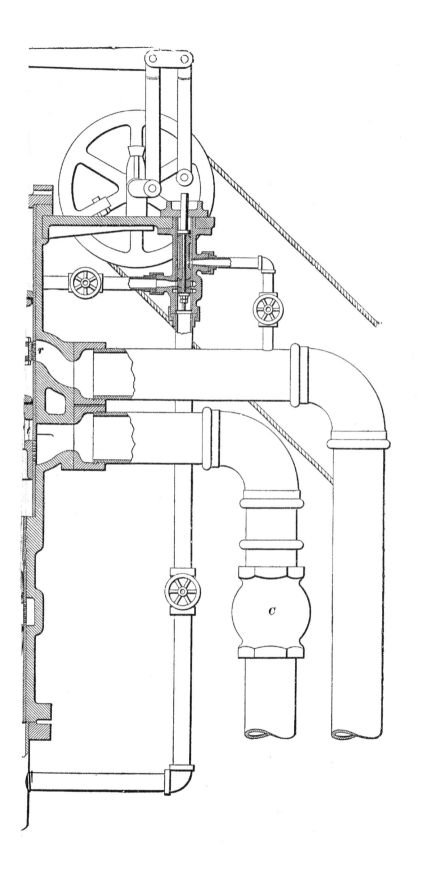

from each of the two sets a rope passes to an overhead sheave and thence to the car. The fixed sheaves are arranged below the traveling sheaves, the latter being attached to a crosshead carried on the piston rods and guided on rails *R, R*. The piston moves *up* for the ascent of the car and *down* for the descent, so that the piston rods are in compression. Moreover, the piston moving in the same direction as the car cannot, as in the case of the previously described vertical machines, be utilized as a counterweight, but must itself be counterbalanced, which is done in the manner shown in the illustration, *W, W* being the weights. The controlling valve is much of the same construction as that shown in Fig. 5.

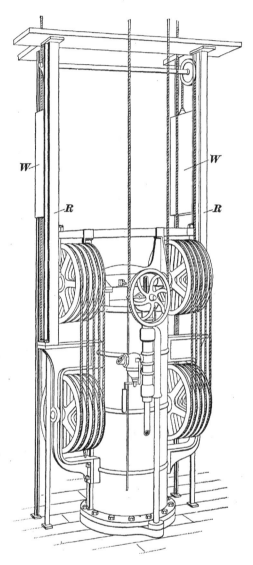

Fig. 12.

27. While in high-ratio vertical elevators of the kind shown in Fig. 12 the circulation of water is dispensed with, owing to the small head of the water, it becomes entirely dispensable when the cylinder is placed horizontally. All horizontal hydraulic piston elevators are, therefore, based on the non-circulating system.

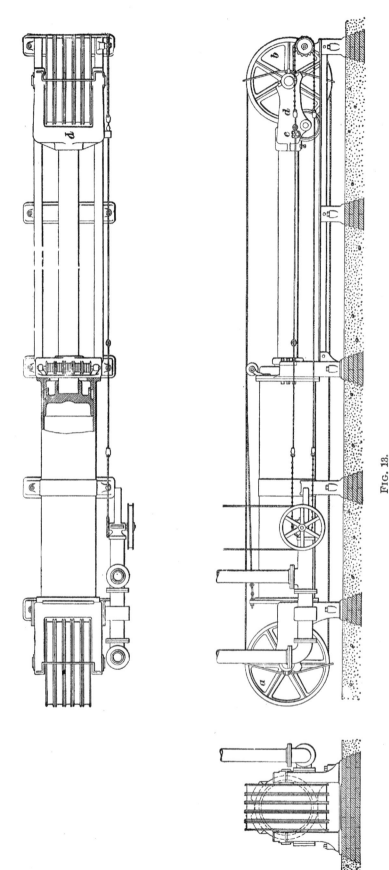

Fig. 13.

HORIZONTAL HYDRAULIC PISTON ELEVATORS.

28. Advantages.—Although the floor space occupied by a vertical elevator cylinder is comparatively small, this floor space is required on each floor of the building, and where there are a number of elevators, the aggregate necessary space amounts to more than can in many instances be conveniently spared. Moreover, it becomes necessary, in case of a battery of elevators, to provide a separate well for the cylinders. Again, the long, upright cylinders so placed in a comparatively narrow well are inaccessible for the greater portion of their length. For these reasons preference is given to the horizontal type of elevator when there is sufficient floor space more available in the basement of the building than on the floors above. But under the most favorable conditions, floor space, even in the basement, is always limited, and it is desirable, therefore, to make the cylinders short, which necessitates a high ratio of the transmitting devices. This is generally chosen as 10 : 1. The sheaves in these machines are arranged either so as to put the piston rod in compression or so as to put them in tension.

29. Compression Type.—A simple machine of the compression type, built by Morse, Williams & Co., is shown in Fig. 13. The fixed sheaves are placed at the rear end of the cylinder and the hoisting rope is carried above and below the cylinder from the fixed sheaves a to the traveling sheaves b back and forth and is finally led off from the former to the car. The drawing calls for but little explanation. The controlling device consists of the three-way valve illustrated in Fig. 5; the motor safeties are limit-stop buttons carried on an endless chain or rope and actuated at the extreme positions of the piston by an arm or projection c on the crosshead d. The endless chain runs over a sprocket wheel fastened to the shipper-sheave shaft.

30. Tension Type.—The general arrangement of the tension type of horizontal hydraulic machines is shown in Fig. 14. Both the fixed and the traveling sheaves are

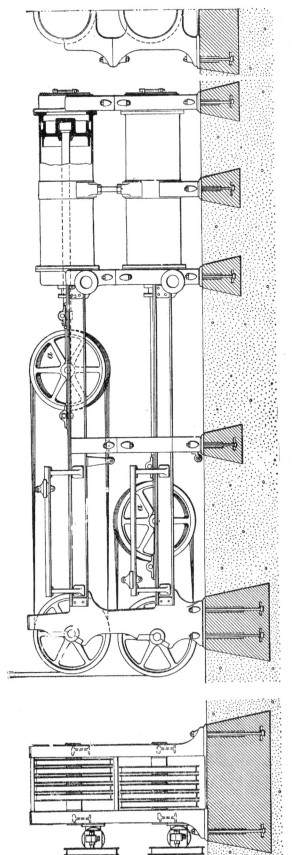

Fig. 14.

located at the front end of the cylinder. It will be noticed that the traveling sheaves a, a are mounted in the crosshead at an angle to the horizontal plane. This is necessary in order that the ropes shall not "ride" off the grooves when the two sets of sheaves come close together at the end of the stroke. This precaution is deemed unnecessary in the compression type, the sheaves being always apart a distance greater than the length of the cylinders.

There are several advantages in the tension type of machines: (1) The piston rods can be considerably smaller. (2) The distance between the fixed and traveling sheaves is smaller, being only about one-half as long as that in the compression type; this is an item of importance when the fact is taken into consideration that teetering of the car is often due to the whipping of the ropes in horizontal machines, which action increases as the distance between the sheaves becomes greater. This action, by the way, is absent in vertical elevators. The whipping of the ropes is reduced as much as possible by supporting rollers shown in Figs. 13 and 14. In the tension type these rollers are supported on a shaft that again rests on guide shoes traveling on rails.

31. The compression type of horizontal elevators has the advantage that no stuffingbox is needed for the piston rod, the water entering behind the piston only. The front end of the cylinder generally has a simple yoke through which the rod passes.

When there is more than one elevator in a building, the cylinders are preferably mounted in pairs on top of each other; such a pair is then called a **double-deck machine,** and this arrangement is shown in Fig. 14.

32. Fast-Service Compression-Type Elevator.—Fig. 15 is an illustration of an elevator machine of the compression type built by the Otis Elevator Company, of Chicago (formerly the Crane Elevator Company). This machine is intended for fast passenger service and is therefore

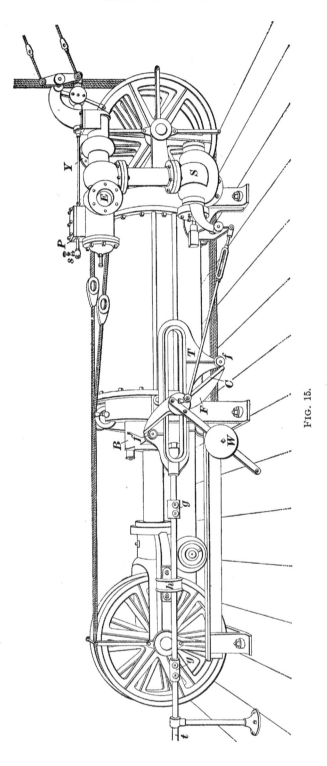

Fig. 15.

fitted with a pilot valve *P*, involving the same principles as the Otis valve described in Arts. **19** and **20**, and has an automatic stop-valve *S*.

33. The pilot valve, main valve, and stop-valve are shown in detail in Fig. 16. The pilot, or auxiliary, valve is a slide

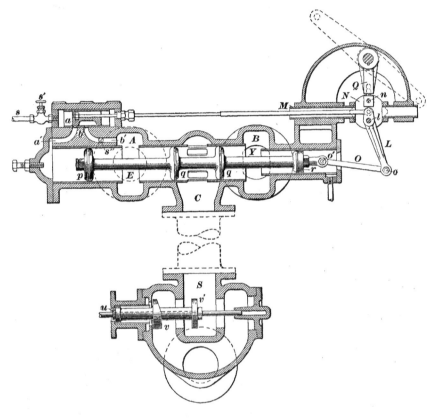

FIG. 16.

valve; its seat has two ports *a* and *b* opening into passages *a'* and *b'* of the main-valve casing. The passage *a'* leads into a space of the main valve behind a piston *p*, while the passage *b'* communicates with a chamber *A* in front of that piston, which chamber, in turn, is connected with the exhaust pipe *E*, Figs. 15 and 16. Of the two other chambers *B* and *C* of the main-valve casing, *B* is connected to the supply pipe *Y*, and *C* to the cylinder by way of the valve S. The main valve consists of two single pistons and one double piston: the piston p, already mentioned, the double piston q, and the piston r, the latter being of smaller diameter than the others. The valve chest of the pilot slide valve is connected with the supply by a pipe s.

The operation of the valves is as follows: In the position shown the valves are at rest, the port a being closed and the pressure on the piston q towards the left (due to the difference in area of q and r), thus acting against a body of water confined in the space behind the piston p. If the pilot valve is moved towards the right, it uncovers the port a and water under pressure enters the space behind the piston p; the area of this piston being greater than the difference of the areas of q and r, it moves towards the right, thus connecting the chambers A and C; that is, connecting the cylinder with the exhaust, and hence the elevator car moves down. If the pilot valve is moved from its neutral position and to the left, the passages a' and b' are connected; that is, the space behind the piston p is put into communication with the exhaust. The excess pressure due to the difference of areas of q and r then causes the pistons to move to the left, opening communication between the chambers B and C; that is, it connects the cylinder with the supply, and hence the elevator car moves up. The speed with which the main valve responds to the pilot valve is regulated by the valve s' in the supply pipe s on one hand and furthermore by a screw s'' that can be made to enter more or less into the passage b' by turning it from the outside.

For the same reason that was given in connection with the Otis pilot valve, Art. **19,** the pilot valve must return automatically to its neutral position. The mechanism that accomplishes this is similar to that used in the Otis valve. The valve stem of the pilot valve is connected to the short arm of a two-armed lever L, which is pivoted at l to the central double disk-shaped piece N of a sliding sleeve M. The long arm of the lever L is connected by means of a link O to the stem of the main valve. The central piece N is connected at n with a one-arm lever Q, the shaft of which is operated by a lever or sheave actuated by a shipper rope from the car. When the lever Q is thrown to the left, the sleeve M moves to the left, carrying the lever L and the pilot-valve stem with it, the point o' at which the link O connects with the main-valve stem being the pivotal point of the motion. As soon as the main valve commences to move to the left, that is, after the pilot valve is set by the shipper rope, the point l becomes the pivotal point, and the pilot valve is pulled back to its original position. Similar action takes place when the lever Q is moved towards the right.

34. The action of the automatic stop-valve S is as follows: The valve has three pistons v, v', and u, of which the first two serve to close the circular openings leading from the inlet to the outlet. The piston u is at all times actuated by whatever pressure there is on the cylinder, forcing it to the left and thus keeping the circular openings referred to open. The valve stem is connected by a lever and rod to a cam F, Fig. 15, pivoted to the frame of the machine. This cam is ordinarily held, as shown in the illustration, between two rollers f and f' by means of a weight W attached to it. The rollers f and f' are placed on a movable frame T guided horizontally as shown and called the **tappet**. On the guide rod t of this tappet are fastened the limit-stop buttons g, g to the right and left, respectively, of a projection, or arm, h on the crosshead of the traveling sheaves. In either of the extreme positions of the crosshead, the arm h comes in contact with one of the buttons, pushing the tappet T and thus operating the stop-valve and shutting off the communication between the main valve and cylinder.

The valves v and v', Fig. 16, are not fitted very closely, so that there is a certain small amount of leakage, which enables the valve to start back slowly as soon as the pilot valve and main valve are reversed; as soon as the arm h, Fig. 15, leaves either of the buttons g and g, the weight W causes the valve to open quickly and wide. In case the leakage around the valves v and v', Fig. 16, proves too slight, a small direct pipe connection (not shown) is made between the middle chamber C of the main valve and the top of the cylinder at the closed end. This allows a small quantity of water, which is regulated by a stop-valve, to enter or leave the cylinder independently of the automatic valve S when the pilot valve is reversed so as to give the valve S the start. This pipe connection also serves the purpose of permitting the escape of air that may have accumulated in the cylinder.

35. The hydraulic elevators described are by no means the only ones that are made or that are in operation. They are typical constructions, however, and a person will, if their principles are clearly understood, readily comprehend other designs as well.

PUMPS, TANKS, PIPES, AND FIXTURES.

GENERAL ARRANGEMENT.

36. In cases where a natural water supply or a street main with sufficient pressure is available, the elevator may be directly connected with it. Such cases are rare, however, and therefore a pumping plant is almost always included in an elevator installation. This pumping plant consists usually of one or more pumps, a pressure tank, and a discharge tank suitably connected by piping provided with the necessary valves and other fixtures.

37. A typical installation of an hydraulic elevator is shown in Fig. 17. The pump P takes the water from the discharge tank D and forces it into the pressure tank T, whence it enters the elevator cylinder C through the supply pipe S. It leaves the elevator cylinder through the discharge pipe t, which carries it back to the discharge tank.

§ 39 ELEVATORS. 33

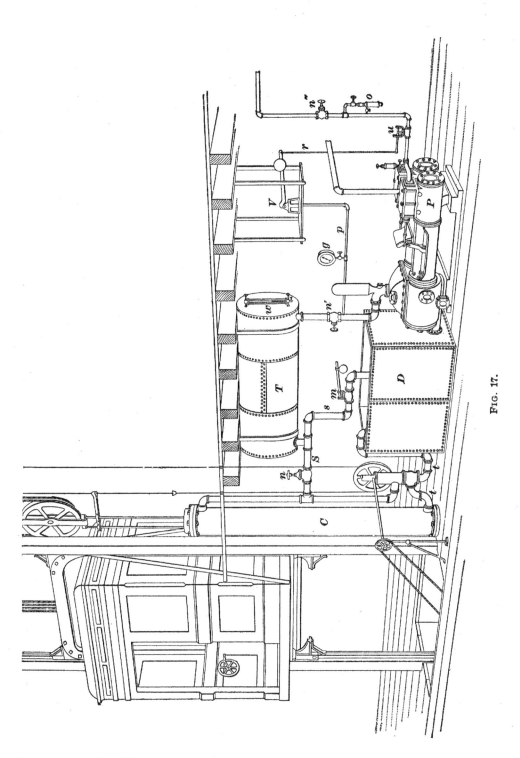

Fig. 17.

The water is thus used over and over again; this is an important item where water rates are high, as is the case in most cities and towns.

PUMPS.

38. Since, with the usual arrangement of pumps, cylinders, and tanks, the pump may work continually while the cylinder takes a quantity of water out of the pressure tank only for every other (the up) trip of the car, the pump need be only large enough to supply the average quantity of water per unit of time, supposing the cars to be running continuously up and down. Since there is more or less interruption of traffic, the pumps will generally even then supply more water than is necessary and will have to be stopped and started frequently. For such intermittent service duplex steam pumps or electric pumps are most suitable and are, hence, generally used, although geared pumps, belt-driven pumps, and gas-engine power pumps are occasionally met with.

TANKS.

39. Open tanks, formerly installed in great numbers on the roofs of buildings to furnish the necessary head, are gradually disappearing, and the closed pressure tank, as T, in Fig. 17, placed in the engine room, takes its place almost exclusively. Such closed pressure tanks are often placed at the top of the building also, thus utilizing both the natural head and the air pressure. In such a tank the required water pressure is obtained by having the tank partly filled with air and compressing the same by pumping in the water, so that it is really air pressure that gives to the water the necessary head. By leakage as well as by absorption in the water the quantity of air in the tank gradually grows less and must be renewed. In the smaller installations, such as is shown in Fig. 17, the necessary air supply is obtained through a vent in the suction pipe of the water pump; in large installations separate air pumps are provided for the purpose.

ACCUMULATORS.

40. The pressure used in ordinary closed-tank installations ranges generally between 125 and 150 pounds per square inch. In some cases for high buildings these pressures have to be exceeded, and then hydraulic accumulators are installed instead of the pressure tanks. These high-pressure installations require also different designs of cylinders and other parts of the plant, but since there are but comparatively few of these installations in operation we shall forego treating them in detail.

AUTOMATIC STOPPING AND STARTING DEVICES FOR PUMPS.

41. Kinds of Starting Devices.—The stopping and starting of the pumps are effected automatically by various devices. In one kind of these devices the height of the water in the tank is made use of by means of a float to operate the steam valve of a steam pump or the switch and rheostat of an electric pump; in another kind, the pressure in the tank is utilized to do the same thing by means of a pressure valve. Floats are used only with open gravity tanks.

42. Pressure-Regulated Starting Valve.—A device of the second of the above-named classes is shown in Fig. 17 at V. It consists of a pressure valve of much the same construction as a steam-boiler safety valve. It is connected to the pump discharge pipe or directly to the pressure tank by a small pipe p, into which is inserted a pressure gauge g. The weighted lever of the valve V is connected to the throttle valve u of the steam pump by a rod r in such a manner that the throttle valve shuts off steam when the weight on the lever of the valve V is balanced by the required water pressure in the pressure tank, and opens to admit steam when the pressure falls below the required amount. A sight-feed oil cup o is generally placed in the steam pipe in advance of the throttle valve u, in order to insure the proper working of the same and to prevent it from sticking.

43. Ford Regulating Valve.—In the device shown in Fig. 18, the two valves V and u spoken of in the previous article are combined into one. This device, which is largely used in elevator work and is manufactured by Thomas P. Ford, of New York, consists of a spring-actuated steam valve U and a water piston V moving in a little cylinder W under the influence of the water pressure. It is easy to see that by adjusting the spring properly the steam valve can be made to close when the water pressure on the piston V exceeds a certain required amount. The regulating valve should be placed in the steam supply pipe in a vertical position between the steam chest and an ordinary throttle valve. The oil cup should be placed so as to allow the oil to pass through the regulating valve. The pipe connecting the pressure tank with the pressure cylinder of the regulating valve should be provided with a globe valve and a union next to the valve, in order that the cap may easily be removed for repacking the piston V. A drip pipe should be connected with the bottom of the cylinder W.

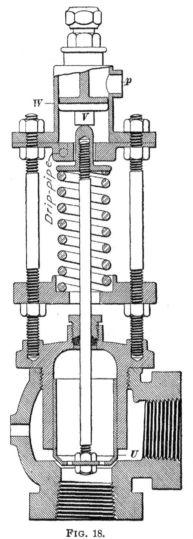

Fig. 18.

44. Ford Rheostat Regulator.—A device much used in connection with electric pumps and manufactured by

Thomas P. Ford is illustrated in Fig. 19. The purpose of this apparatus is to obtain a comparatively large movement, which is necessary for operating the switch and rheostat of the electric motor.

As in the apparatus shown in Fig. 18, the pressure pipe is connected to a small cylinder W in which works a piston V

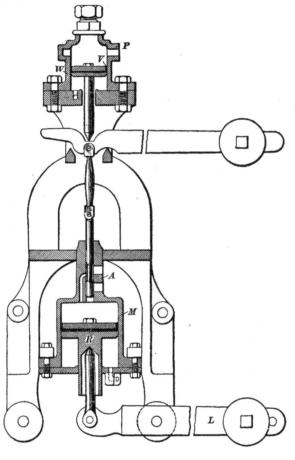

FIG. 19.

against a weighted lever. This lever is, however, not connected directly to the stopping and starting arrangement, but to the piston of an auxiliary hydraulic valve A. This valve has an inlet connected to some constant water supply of moderate pressure (not less than 35 pounds per square inch) and a discharge outlet. When the pressure in the tank falls below the required amount, the piston V rises and carries with it the piston of the auxiliary valve; water is then admitted into the cylinder M of the main valve, causing the piston R therein to be forced down and the outward end of a long double-armed lever L attached thereto to be forced up. This lever is also weighted and to it is attached the lever of a motor starting box. As soon as the pressure in the tank increases, the piston V moves down; by this movement the cylinder M is put in communication with the discharge, whereupon the main-valve piston moves up and the end of the lever L down by virtue of the weight attached to it. It is recommended in connecting up this valve to have the water from the constant supply go through a mud-drum placed near the regulator before entering the same.

45. Mason Elevator Pump-Pressure Regulator.—Fig. 20 shows a regulating device much used in elevator work. Referring to the illustration, the operation is as follows: Steam from the boiler enters the regulator at the point marked "inlet" and passes through into the pump, which continues in motion until the required water pressure is obtained in the system, which acts through a $\frac{1}{4}$-inch pipe connected at a and upon the diaphragm D. This diaphragm is raised by the excess water pressure and carries with it the weighted lever L, opening the auxiliary valve A and admitting the water pressure from the connection b to the top of the piston P, at the same time opening the exhaust port under the piston P, thus allowing the water under this piston to escape through the passage a' shown in dotted lines into the drip pipe d, thereby pushing down the piston, which closes the steam valve and stops the pump.

As soon as the pressure in the system is slightly reduced, the lever L, on account of the reduced pressure under the diaphragm, is forced down by the weights W, carrying with it the auxiliary valve A and thus opening the exhaust from the top of the piston, and at the same time admitting the water pressure under this piston, which is now forced up and opens the steam valve, again starting the pump. This action is repeated as often as the pressure rises above and falls below the required amount.

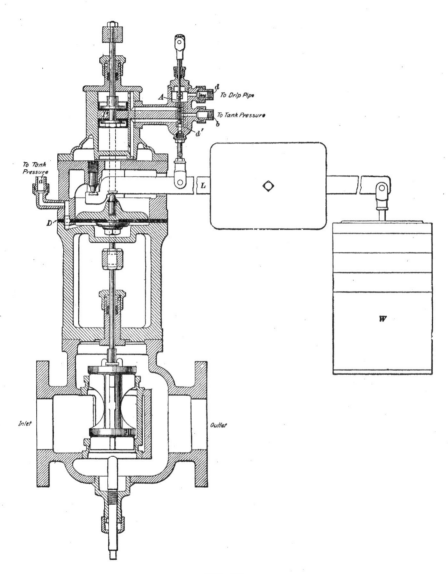

FIG. 20.

46. The Mason regulator is easily adapted for use in connection with a switch and rheostat for regulating electrically driven pumps. Fig. 21 shows such an arrangement comprising a solenoid rheostat and snap switch as made by the Elektron Manufacturing Company, a Perret motor, and a Mason regulator.

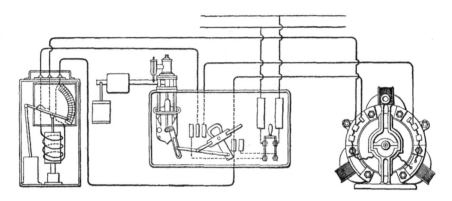

Fig. 21.

BY-PASS VALVE.

47. When the elevator service is quite continuous and regular it proves advantageous in many cases, especially with pumps driven electrically or by gas engines, to have the pump run continually and thus to do away with the more or less complicated automatic-valve switch and rheostat arrangements. In such cases a **by-pass valve** is installed near the pump, which opens communication between the delivery and suction pipes of the pumps whenever the pressure in the tank becomes excessive. By elevator men, such an arrangement is called a **closed system**.

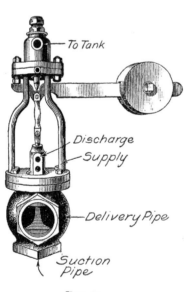

Fig. 22.

Fig. 22 is an illustration of the Ford by-pass valve. Its construction is similar to that of the regulator described in Art. **44**, and it is connected up in the same manner, a mud-drum being preferably placed near the valve to free the water from any impurities before it enters the auxiliary valve.

SAFETY VALVE.

48. To provide for the emergency, should the regulating devices described in the previous articles stick, and should an excessive pressure accumulate in the tank, a pipe *s* fitted with a **safety valve** *m* (see Fig. 17) and leading from the pressure tank to the discharge tank is generally provided.

WATER GAUGE AND VALVES.

49. Besides the fixtures already mentioned, there is provided a water gauge *w* on the pressure tank and various globe valves *n*, *n'*, and *n''*, Fig. 17, which are used in starting and stopping the plant.

OPERATION AND MAINTENANCE OF HYDRAULIC ELEVATOR PLANTS.

50. Water.—The water to be used in hydraulic elevators should be clear and free from sediment. It should enter the system through a strainer, so as to exclude all foreign matter likely to damage the valves and pistons. The water should be changed at least every three months and the whole system should then be cleaned by washing and flushing. This requires closing down the plant completely.

51. Starting Up and Running.—With all parts supposedly in good working order, joints tight, stuffingboxes, pistons, and valves properly packed, guides, sheaves, and other moving parts well oiled, start the pumps and partially fill the pressure tank; in doing so, the air in the tank will be compressed, but there will not be sufficient air in the tank to give the required pressure. Therefore, when the tank is about half full of water, open the air vent in the suction pipe of the pump, thus introducing air with the water until the proper pressure is reached, when the gauge shows about one-third of air and two-thirds of water, this being the proportion upon which tanks are generally based to amply supply the necessary amount of water for the cylinders. When an extra air pump is provided, fill the tank two-thirds full of water and supply the air pressure afterwards. The water level indicated above should be carefully maintained by the engineer in charge during the operation of the plant by opening the vent in the suction pipe of the pump occasionally or starting the air pumps, respectively, whenever the water level rises higher through loss of air by leakage or absorption. It is better to have a little more air in the tank than too little, since too small an air volume is apt to cause considerable fluctuation of the pressure during each stroke of the elevator piston.

After the necessary pressure has been reached, slowly open the stop-valve between the tank and the controlling valve, which stop-valve is generally and preferably a gate valve. Next, slowly open the controlling valve, *all air valves or cocks having been previously opened*, to allow the air contained in the cylinder to escape; the air cocks are shown at a, Fig. 6, and at b, Fig. 8. For the first filling of the cylinder, the controlling valve must be set for *going up*. After all the air is expelled, which can be ascertained by water running from the air cock into the funnel of the drip pipe m, Figs. 6 and 8, close the air cock. The elevator is now ready for running.

52. Absorption and Discharge of Air.—As already mentioned, the air will be absorbed by the water to a certain extent. This air frees itself in the cylinder and may form a cushion. It is, therefore, occasionally necessary to remove the air. In vertical-cylinder (circulating) systems such an air cushion can form above and below the piston. Air below the piston is automatically removed in the Otis vertical elevator by a piston air valve c, Fig. 6, provided for the purpose, which lets the air into the space above the piston, whence it can be removed through the air cock *a*. When there is air in the cylinder, this will cause the car to spring up and down in stopping. When the quantity of air is small, it can generally be let out by opening the air cock and running a few trips. This should, therefore, be done occasionally. If there is much air in the cylinder the car must be run to the top and the controlling valve set for *going down*. While the car and valve are in this position, open the air cock and allow the air to escape. This may have to be repeated several times before the air is all removed. If the absorption of the air by the water is found to be considerable, it may effectually be prevented by the introduction into the tank of a layer of heavy oil about 4 inches thick. This expedient will, however, have to be resorted to but seldom.

53. Settling of Car.—After all the air is removed close the air cock, as otherwise the car will settle, that is, slowly creep down at the landings. If the air cock is properly closed and the car still shows a tendency to settle, the cause is probably that the piston or valve is leaking and needs repacking. Another cause for settling may be that the piston air valve *c*, Fig. 6, does not properly seat.

54. Groaning Noise in the Cylinder.—If a groaning noise is heard, it may be taken as a sign either that the two piston rods (in the vertical type) are not drawing alike or that the piston packing is worn out and needs renewal. If it is believed that the fault lies with the rods, this may be ascertained by trying to turn the rods with the hands; if one of them will turn, it needs tightening up. If the packing is at fault, the car will settle.

55. Stretching of Cables.—The cables should not be allowed to stretch enough to prevent the car from reaching the top landing, because of the danger of the piston striking the bottom cylinder head.

56. Hand Cable, Limit-Stop Buttons, Back-Stop Buttons.—The hand cable, or shipper rope, as we have called it, should be properly adjusted, neither too tight nor too loose. The limit-stop buttons should be so adjusted that the car will stop at a few inches beyond either extreme landing and before the piston strikes the head of the cylinder. The back-stop buttons should be so adjusted that the valve cannot be opened either way more than to give the car the required speed. In the case of auxiliary, or pilot, valves, the stops on the shipper sheave serve instead of the back-stop buttons.

57. Lubrication.—The plungers in plunger elevators should be kept well greased and clean. A good way to clean and grease the plunger, suggested by the Plunger Elevator Company in connection with their "elevator grease," is to stand at the bottom floor and to run the elevator slowly up while *wiping the plunger dry*. On running the car down again, cover the plunger with a thin coat of grease, rubbing it on and spreading it even with the hands. The plunger should be dry when the grease is applied; otherwise the grease will not stick. The inside of the cylinder should be lubricated about every two weeks with cylinder oil. Oil cups are generally provided for this purpose. The Otis Elevator Company, of Chicago (Crane Elevator Company), say the following in regard to lubrication: "The most effectual method of lubricating the internal parts of hydraulic-elevator plants, where pumps and tanks are used, is to carry the exhaust-steam drips from the foot of the pump-exhaust pipe to the discharge tank, thus saving the distilled water and cylinder oil. This system is invaluable when water holding minerals in solution is used, as these minerals greatly increase corrosion."

Horizontal machines operated by city pressure are best lubricated with a heavy grease, applied either mechanically or by means of a piece of waste on the end of a pole. The former method serves as a constant lubricator, while in the latter case greasing is often neglected and, in consequence, packing lasts but a short time.

Mr. Ford recommends as a lubricant for his valves, described in Arts. **43, 44,** and **47,** common soap applied once a month.

58. Bushing Sheaves.—If the traveling-sheave bushing is worn so that the sheave binds or if the bushing is nearly worn through, turn it half around and thus obtain a new bearing. If it has been turned before, put in a new bushing.

59. Precautions Against Freezing.—Precaution must be taken against the water freezing in any part of the system. If the cylinder and connections must be located in an exposed place, they should be protected against frost by building an air-tight box, open at the bottom, around them; a small gas jet should be kept burning at the bottom, or when steam is available a coil should be placed near the cylinder. Plunger-elevator cylinders are exempt from the danger of freezing. Supply pipes outside of the building are best protected by burying them in the ground below the freezing line, say 6 feet. If this cannot be done, they should be covered with non-conducting material, the same as is used for steam pipes. If in cold weather the elevator service is to be stopped for any length of time the water must be drained off, care being taken that this is done thoroughly. This applies especially to small pipe connections for drips, vents, etc., which should be free from bends, loops, or sags in which water may be left to freeze after the system has been drained.

60. Closing Down Hydraulic Elevators.—We will imagine that for some purpose, as prevention of freezing, change of water, etc., the plant is to be closed down. After removing the lower limit-stop button, run the car slowly to the bottom. Next shut off the supply by closing the valve provided for the purpose in the supply pipe, as the valve n in Fig. 17. In the plunger type of elevator machine, the valve and connections only are thus drained, the cylinder remaining full of water around the plunger, which, however, does no harm, since being far underground the water will not freeze. In the horizontal machines, running the car down and closing the supply leaves both cylinder and valve free of water. In the vertical (circulating) machine, however, the cylinder and circulating pipe are still full of water when the car is down and must be drained. For this purpose, open the air cock and the drain-pipe valve d, Figs. 6 and 8. Throw the valve for *going up* to empty the cylinder through the discharge pipe. Next throw the valve for *going down* to empty the circulating pipe through the drain pipe. After all water is drained off, grease the cylinder with heavy grease if the machine is of the horizontal type, and grease the piston rods if of the vertical type.

61. Packing Plunger and Piston Rods and Stuffingboxes.—Stuffingboxes that must be repacked from time to time occur in the plunger type, the vertical type, and the horizontal tension type of hydraulic elevators. For repacking the stuffingboxes, it is neither necessary nor expedient to drain the system.

For packing plunger stuffingboxes, run up the car sufficiently to be enabled to work conveniently in the pit, shut the three-way controlling valve and the supply stop-valve between the tank and the controlling valve. Block up the car, then remove the gland of the stuffingbox and renew the packing; replace the gland, screwing up the bolts just tight enough to prevent leaking, open the supply stop-valve and then slowly the controlling valve, setting it for *going up*. Remove the blocking.

62. Various materials are used for packing plunger stuffingboxes. For the smaller sizes, such as sidewalk-elevator plungers, fibrous packing, such as hemp, flax, or cotton, is used exclusively. For large plungers cup leathers are probably the best packing. But since the cup-leather ring must be split open in order to introduce it into the box, much of its value is impaired; therefore, fibrous packing is much used.

63. To retain the cup-leather principle and at the same time to avoid the objection to the butt joint, multiple cup leathers may be used. Fig. 23 shows a plan that is said to have proved very satisfactory. The packing consists of split leather rings, or even of ring sections, of **V**-shaped cross-section. The edges of these rings are cut down sharp, in consequence of which they act in much the same manner as cup leathers. The single sections are, of course, intro-

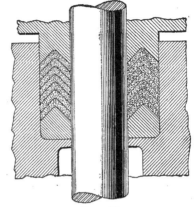

Fig. 23.

duced so as to break joints. This kind of packing is **very** tight, but is likely to create a great deal of friction.

64. A much better arrangement is shown in Fig. 24. This packing, known as **Wright's elevator packing,** con-

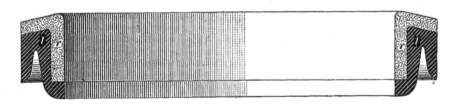

Fig. 24.

sists of a split rubber ring l of cup-shaped cross-section and a split leather ring r of **L**-shaped section. Both rings are placed in the stuffingbox so as to break joints.

65. For packing piston-rod stuffingboxes, close the supply stop-valve and open the air cock to make sure that there is no pressure in the cylinder; remove the followers and glands of the stuffingboxes and renew the packing. Screw down the followers only tight enough to prevent leaking. Fibrous packing is used exclusively.

66. Packing Vertical Cylinder Pistons. — In some designs of vertical elevators the piston can only be packed from the top, as in the elevators shown in Figs. 5 and 8. In others, provision is made for packing the piston either from the top or bottom, as in the Otis elevator shown in Fig. 6. In others, again, the piston can be packed only from the bottom, as in the elevator shown in Fig. 11.

67. To pack a vertical cylinder piston from the top, run the car to the bottom and close the stop-valve in the supply pipe. Open the air cock at the head of the cylinder and also keep open the valve in the drain pipe from the side of the cylinder long enough to drain the water in the cylinder down to the level of the top of the piston. Now remove the top head of the cylinder, slipping it up the piston rod out of the way and fastening it there. If the piston is not near enough to the top of the cylinder to be accessible, attach a rope or small tackle to the *main cables* (not the counterbalance cables) a few feet above the car and draw them down sufficiently to bring the piston within reach. Remove the bolts in the piston follower by means of a socket wrench. Mark the exact position of the piston follower before removing it, so that there will be no difficulty in replacing it.

In the elevators shown in Figs. 5 and 8 fibrous hemp packing is used. In the design shown in Fig. 6, a combination of cup-leather and duck packing is used. On removing the follower of this piston, a leather cup l is found turned upwards, with coils u of $\frac{5}{8}$-inch square duck packing on the outside. This duck packing should be removed and the dirt cleaned out; also clean out the holes in the piston through which the water acts on the cup. If the leather cup is in good condition, replace it and on the outside place three new coils of $\frac{5}{8}$-inch square duck packing, being careful that they break joints and also that the thickness of the three coils up and down does not fill the space by $\frac{1}{4}$ inch, as in such a case the water might swell the packing sufficiently to cramp it in this space, thus destroying its power to expand. If too tight, strip off a few thicknesses of canvas.

Replace the piston follower and let the piston down to its right position. Replace the cylinder head and gradually open the gate valve in the supply pipe, first being sure that the operating valve is on the center. As soon as the air has escaped, close the air cock and the elevator is ready to run.

68. To pack vertical-cylinder pistons from the bottom, remove the top limit-stop button and run the car up until the piston strikes the bottom head of the cylinder. Secure the car in this position by passing a strong rope under the girdle, or crosshead, and over the sheave timbers. When secured, close the gate valve in the supply pipe, open the air cock at the head of the cylinder, and throw the controlling valve for the car to *go up*. Also open the valve in the drain pipe from the side of the cylinder and from the lower head of the cylinder, thus allowing the water to drain out of the cylinder. When the cylinder is empty, throw the valve for the car *to descend*, in order to drain the water from the circulating pipe.

In cases of tank pressure, where the level of water in the lower tank is above the bottom of the cylinder, the gate valve in the discharge pipe will have to be closed as soon as the water in the cylinder is on the level with that in the tank, allowing the rest to pass through the drain pipe to the sewer. When all water is drained off, proceed as directed in the previous article in renewing the packing. To refill the cylinder after packing, close the valves in the drain pipes, leave open the air cock at the head of the cylinder, leave the controlling valve in the position to descend, and open the gate valve in the discharge. Slowly open the gate valve in the supply pipe, allowing the cylinder to fill gradually and the air to escape at the head of the cylinder. When the cylinder is full of water, close the air cock and put the controlling valve on the center. The car can then be untied, the limit-stop button reset, and the elevator is ready to use.

69. Packing Horizontal Hydraulic Elevators.—In a compression-type elevator run the car to within 1 foot of the extreme top and secure it to the overhead beams with a chain or rope. Close the gate valves in the supply and discharge pipes and open the air cock and valve in the drain pipe, emptying the cylinder. Remove the buffer across the front (open) end of the cylinder and slide it along the piston rod out of the way. Remove the follower of the piston. With a hooked piece of wire remove the old packing. Raise the piston head until it is in the center of the cylinder. If the cylinder is found to be in good condition, cut off four rings of square lubricated fibrous packing 9 inches longer than the circumference of the cylinder. Place the two ends of a ring together and form tucks with the balance. Force in these tucks one at a time with a hardwood stick until all are level against the head. Proceed in the same manner with the remainder of the packing. Arrange the packing so that the joints in the different rings do not come together.

If the cylinder is badly worn, use square pure-rubber packing for the first and last ring, and make these but 1 inch larger than the circumference of the cylinder. This rubber insures a backing for the fibrous packing. After putting the packing in position, replace the follower and screw on the nuts with the fingers until the follower is close to the packing. On two of the studs opposite each other will be found jam nuts. Set these out against the follower and tighten with a wrench. Replace the buffer. Close the drain valve and set the controlling valve for going up. Open the gate valve in the supply pipe and fill the cylinder. When the cylinder is filled, close the air cock. As the car in the first place was not at the extreme top, the pressure in the cylinder will run the piston head against the buffer and the car will ascend to the extreme top. The fastenings may then be removed. Throw the controlling valve on the center and open the discharge. The elevator is then ready to descend. Do not make any trips until the cylinder is thoroughly greased. Continue greasing twice a week.

In the course of time, leaks will occur in the cylinder. Loosen the jam nuts back of the follower and set up the nuts on the studs equally until the leak is stopped. Then retighten the jam nuts.

70. In the tension type of horizontal hydraulic elevator, the procedure is exactly the same, with the exception that there is no buffer to be removed, the open end of the cylinder being at the back.

71. Packing the Controlling Valves.—Run the car to the bottom, close the supply valve, and drain the system as previously described. When the water is all drained off take off the cap. After marking the exact position of the various parts in relation to one another, remove the valve proper and renew the packings, placing the new ones in the same position as the old ones. Before refilling the cylinder close the valves in the drain pipes, but leave the air cock at the head of the cylinder open and be careful that the valve is in the position for the car to go down. Gradually open the gate valve in the supply pipe. When the cylinder is filled, close the air cock and open the gate in the discharge pipe.

72. Packing Material.—Fibrous packing is furnished by the trade in the form of a square braided fiber impregnated with a greasy substance. The material used is hemp, flax, or cotton. It is claimed by some that cotton is a more suitable material, being more elastic, softer, and more absorbent for grease. In using it, it is important that it should be well soaked in boiling tallow for several hours to exclude with certainty all air from the pores.

73. Leather for cups should be of the best quality, of an even thickness, free from blemish, and treated with a waterproof dressing. The cups should be of sufficient stiffness to be self-sustaining when passing over the perforated valve lining. Elevator builders generally make and furnish packings to fit their machinery, and it is recommended to get supplies from them. When ordering cups, the pressure of water carried should be specified, as the stiff cups intended for high pressure would not set out against the valve lining when low pressure is used.

ELEVATORS.

(PART 4.)

CAR SAFETIES.

PURPOSE.

1. The term **car safeties** applies to safety devices that in cases of emergency prevent the car from falling unretarded to the bottom of the shaft. All these devices, with the exception of air cushions, consist primarily of catches in the shape of wedges, pawls, etc. that lock the car to the guides. They differ, however, in the means by which these catches are set in operation. In some designs of car safeties, only the breaking of the hoisting cable or cables or its becoming slack through a temporary sticking of the car in its descent will operate the safety catches. In other designs, excessive speed of the car is relied on to operate them.

SLOW-SPEED CAR SAFETIES.

2. Fig. 1 shows the simplest form of a car safety intended to operate at the breaking of the hoisting cable or its becoming slack through a temporary arrest of the car. The hoisting cable is attached to a bolt F, which is free to slide in its hole in d, but has an enlarged head on the bottom through which the curved spring e passes. The lower end c

of the bolt is slotted to receive one end of the bell-crank levers *E, E*, which are pivoted to the uprights of the car. The other ends of the bell-crank levers carry pawls *f, f*, which are spring-actuated and adapted to enter between suitable ratchet teeth on the guides. The pawls are normally held out of engagement with the teeth, the spring *e* being compressed by the load. Should the cable break or become slack for any reason, such as a temporary arrest of the car in its descent, the tension in the spring *e* would be relieved and the pawls would consequently engage the ratchet teeth, preventing the car from falling.

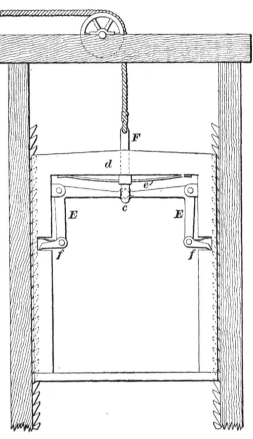

FIG. 1.

3. Pawl-and-ratchet arrangements, such as are shown typically in Fig. 1, are now but seldom used; they are suitable only for slow elevators. The pawl is generally replaced by a wedge that enters between the guides and the guide shoes. Fig. 2 is an example of a car safety of this kind. The cable is attached, as in the previous case, to a spring-actuated bolt or stirrup *F* carrying an **S**-shaped plate *S*, to which the links *L, L* are attached. These, in turn, are connected to levers *E, E*. When the cable breaks, the helical springs surrounding the legs of the stirrup force down the plate *S*, lifting up the outer ends of the levers E, E. These levers, in turn, then press on serrated wedges W, W contained in pockets of the guide shoes in such a manner that ordinarily they remain

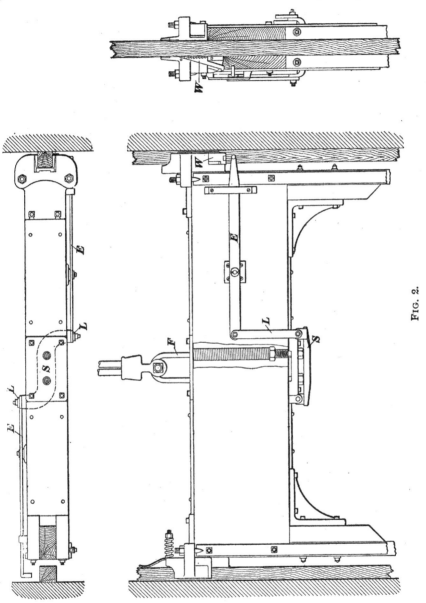

Fig. 2.

by gravity out of contact with the guides. The pressure of the lever ends forces them against the guides and the downward motion of the falling car wedges them tight, the

FIG. 3.

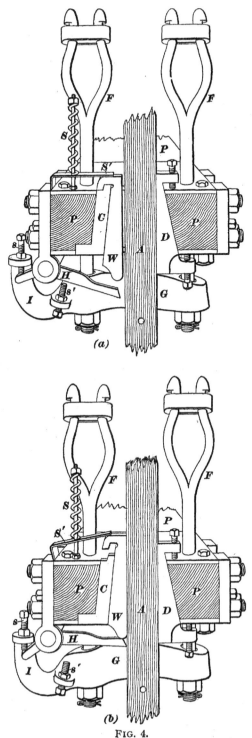

FIG. 4.

serrations or teeth burying themselves into the wood of the guides. When the car is again lifted, the wedges disengage themselves. This safety is regularly used by the A. B. See Manufacturing Company on their freight elevators with wooden guides.

4. It will be noticed in the arrangement shown in Fig. 2 that two cables are used, but in order that the safety mechanism should operate, both cables must break or become slack at the same time.

5. In order to make the safety device respond to the breaking, slacking, or even stretching of one of several cables used, the cables must be independently attached to wedge-operating levers. Figs. 3 and 4 show an arrangement of this kind known as the **Otis gravity-wedge safety**. It consists of a so-called **safety plank** P of hardwood placed under the platform of the car, and into the ends of which are let the guide shoes, each of which consists of a fixed jaw C and an adjustable one D, the latter being very clearly shown in Fig. 4 (a) and (b). Between the fixed jaw and the guide A is inserted a wedge W, which is normally held by gravity in such a position that the shoe can slide freely over the guide. The cables are attached, in the manner shown in Fig. 3, to the **shackle rods** F, Fig. 4, which, in turn, pass through an **equalizing lever** G pivoted in a suitable manner to the safety plank P. From the shackle rods the cables are carried upwards over rollers in a wrought-iron **girdle** B, Fig. 3, at the top of the car. By virtue of the equalizing levers, each of the four cables carries an equal strain, and as long as all cables are equally sound the equalizing levers will remain in their original position, that is, horizontal, as shown in Fig. 4 (a). As soon, however, as one of the cables breaks or even stretches more than its neighbor, the equalizing lever will tilt, as shown in Fig. 4 (b). An arm I of the equalizing lever carries a setscrew s, which is so adjusted that when the lever tilts down it will strike the end of a **finger** H mounted on a shaft under the safety plank. This finger then presses against the wedge W, making it engage the guide. Another setscrew s' is provided on the arm of the equalizing lever and has the same effect on the finger H in case the lever tilts the other way. It is thus seen that in case any of the cables become slack, stretched, or broken, the car will be stopped. The car may then be lifted by the other cables, but it cannot be lowered until the damaged cable is replaced. The spring S acting on the spring plate S' keeps the wedge W in place and prevents it, under normal conditions, from being drawn into engagement by mere sliding contact with the guide.

HIGH-SPEED CAR SAFETIES.

6. Safeties operated by the breakage, slacking, or stretching of a hoisting cable are today not considered sufficient except for very slow-speed elevators. In all high-speed elevators the catches are set into operation by excessive speed of the car, and the most generally adopted plan to effect this is the employment of a centrifugal governor placed either on top of the hoistway or carried on the car, and operated by an endless rope attached to the car or some fixed point. Such an arrangement is often found in addition to safeties to be operated by breaking cables, notably when city ordinances demand the latter.

7. An example of such a safety device is given in the Otis elevator shown in Fig. 3. The finger shaft mentioned in Art. **5** can also be operated by a rope r attached to a lever l, which, in turn, presses on a finger f on the finger shaft. The rope r passes around the pulley of a centrifugal governor G, Fig. 3, on top of the hoistway and an idler at the bottom. The idler is mounted on a crosshead that slides vertically in short guides and is weighted so as to give the rope r the proper tension. The centrifugal governor, by the outward motion of the balls, operates a clutch consisting of two eccentrics g and g', between which the rope r passes and which are geared together so as to grasp and pinch the rope when the balls move out too far, owing to excessive speed. The shaft of the eccentric g' has a crank o connected by a rod to the operating lever of the ball governor G. The eccentric g' is, however, loose on its shaft and has fastened to it an arm a having a stop pin i, against which the crank o strikes at excessive speed of the governor, bringing the eccentrics together so as to just bite the rope r. The continuing motion of the rope then pulls the eccentrics over fully, finishing the grip. The governor thus only starts the gripping of the eccentrics. It will be easily understood that reversing the motion of the car will throw the eccentrics back into their original position. The gripping of the rope causes the descending car to turn the lever l left-handed; this, in turn, rotates the finger shaft through engaging the finger f, and the finger H then operates the wedge W; the guides are thus gripped.

8. A governor-operated safety device used in Otis elevators having steel guides, and for speeds up to 200 ft. per min., is shown, together with the whole car frame, in Fig. 5 and in detail in Fig. 6. It will be noticed from the drawing that the four hoisting cables are not connected in any way

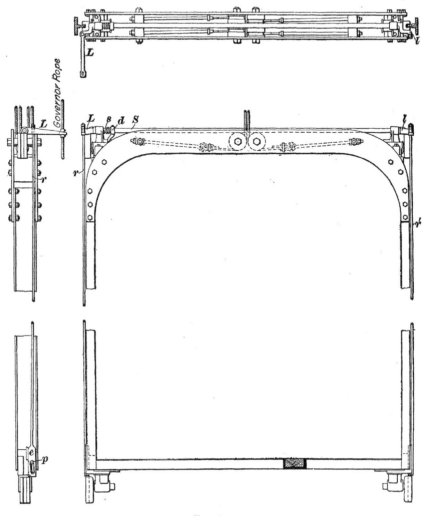

FIG. 5.

with the safety device, but that the latter is solely operated by the governor rope. The rope is attached to a lever L, which is fastened to a shaft S running across the top of the car frame. This shaft is held, normally, in a fixed position by a helical spring s and a stop-collar, or dog, d, resting against the guide-shoe casting. A little nearer the fulcrum of the

§ 40　　　　　　　　ELEVATORS.　　　　　　　　9

lever L a rod r is attached to this lever; and to a separate lever l on the other end of the shaft S a similar rod r' is attached. These rods extend downwards to the safety plank, where they have flattened ends e, Figs. 5 and 6. A slot in each of these flattened ends serves to guide the rods by means of a pin p. On the under side of the flattened end is a shelf f, Fig. 6, that supports a loose roller r, serrated on

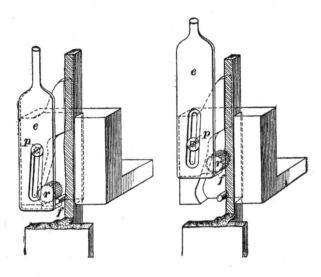

Fig. 6.

its cylindrical surface and contained in a pocket formed in the casting on the end of the safety plank. This pocket is so formed that if the roller r be lifted, it will be wedged in between the back wall of the pocket and the side of the T-shaped guide rail. The operation of this arrangement will be easily understood from the above brief description and Fig. 6, which shows the clamping roller in action and out of action.

9. In some governor-operated safeties the governor is carried under the platform of the car. Fig. 7 shows an arrangement of this kind as built by the A. B. See Manufacturing Company, and is intended for use with steel guides. Fig. 7 (*a*) is a bottom view, while Fig. 7 (*b*) is an

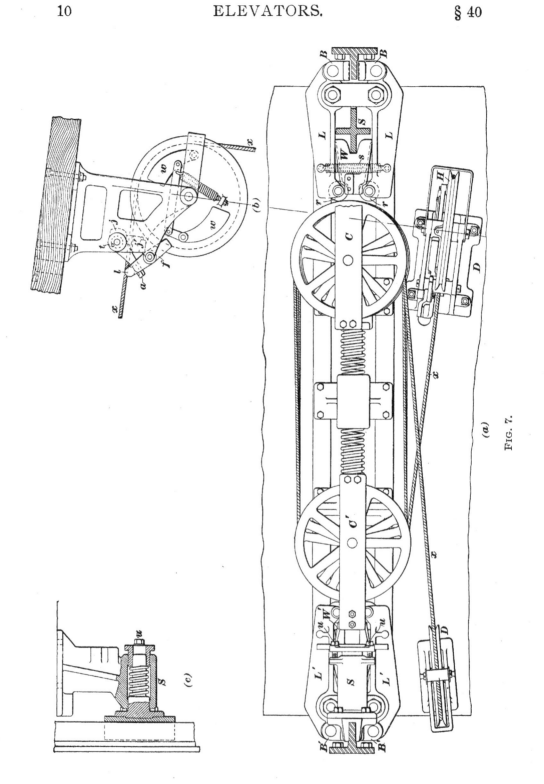

Fig. 7.

elevation of the governor and cable-gripping device. To the guide-shoe castings S, shown in side view in Fig. 7 (c), shown complete at the left of Fig. 7 (a), and at the right with the guide shoe proper and its sleeve removed, are pivoted the levers L, L and L', L'. The short arms of these levers carry grip blocks B, B and B', B', which are intended to close upon the guide rails in case of excessive speed. The long arms of the levers L, L' carry rollers r, r. Each pair of levers is connected by a spring s that normally holds the grip blocks off the guide rail. The governor rope x passes up from the bottom of the elevator shaft over the governor sheave H to the first one of a set of sheaves mounted in a crosshead C', thence to the first one of another similar set of sheaves mounted in a crosshead C, thence back and forth over the other sheaves of these sets, and finally over an idler D up to the top of the hoistway. The crossheads C and C' are properly guided and held by springs a certain extreme distance apart under normal conditions. To the crossheads are bolted cast-iron wedges W, W', which are so designed as to enter between the rollers r, r on the ends of the long arms of the levers L, L', and thus to push the same apart, closing the grip blocks down on the guide rails. These wedges enter between the rollers when the governor rope is arrested by the gripping device on the governor, since then the two sets of sheaves mounted on the crossheads C, C' will be pulled together by the rope shortening between them.

10. The action of the governor will be easily understood from Fig. 7 (b). The governor rope x coming from the governor sheave H passes between two jaws j and j', the former of which is pivoted to the governor frame and is actuated by a helical spring t that gives it a tendency to bear down on the rope against the other jaw j', which is fixed. The movable jaw j has an arm a attached to it, over which hooks the lug l on one end of a double-armed lever or finger f. The other end of the lever f projects into the paths of the governor weights w, w so as to be struck by them when they fly out too much, owing to excessive speed. In this case the lug *l* releases the jaw *j*, the rope *x* is locked to the frame, and the safety is put into action.

SAFETY DRUM.

11. Fig. 8 is a diagram of an arrangement often met with on Otis steam elevators. A so-called safety drum *S* is placed on the same shaft as the overhead sheave *H* for the hoisting rope. Attached to this safety drum are two ropes; one, the **safety rope** *s*, runs down to the levers of a suitable car safety on the car, and the other one, *t*, which is wound the reverse way on the drum, runs down to the hoisting drum; this rope is called the **take-up rope**. When the car is ascending, the take-up rope winds the safety rope on the drum *S*. If the hoisting cable *C* should break, the weight of the car would come on the safety rope and thus throw the car safety into action. The hoisting rope is generally also connected to an independent car safety.

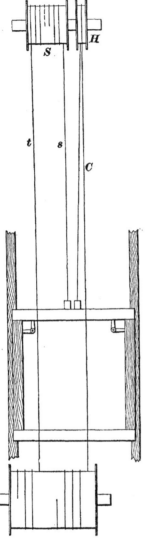

FIG. 8.

12. In connection with the safety drum, a governor-controlled brake is generally used, which, if the hoisting rope should break, insures a gradual fall of the car, thus giving the safety time to act without a sudden shock.

The governor and brake are shown in diagrammatic form in Fig. 9, where *S* is the safety drum, *B* the brake pulley, and *G* a spur gear driving a pinion *P*. From the shaft of this pinion motion is transmitted to the governor spindle by bevel gears, as shown. The sleeve of the governor operates a bell-crank lever L having a projection l, on which is supported, by a hook h, the brake lever W. It is easy to see

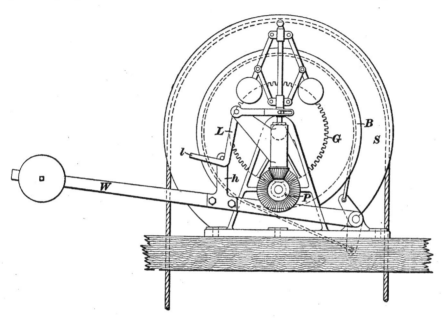

FIG. 9.

that when the governor balls fly out owing to the excessive speed of the car, the arm l will pass from under the hook h, and the weight on the brake lever W will apply the band brake.

13. The different designs of car safeties in actual use are very numerous, but a person understanding the operation of those here described will be able to understand the operation of most of them.

CARE OF CAR SAFETIES AND GUIDES.

14. The importance of keeping car safeties and guides clean and well lubricated, so that they will promptly do their duty when called upon, cannot be emphasized too strongly. Car safeties need adjustment from time to time,

15. When the guide shoes are adjustable, as most of them are, they should be so adjusted that the car will not wabble, but they should not be tight enough to bind on the guide rails. With spring-actuated guide shoes, such as are shown in Fig. 7 (*c*), for instance, the proper adjustment is easily accomplished by manipulating the screw bolts u in the same manner as the bolts of a stuffingbox.

In the Otis wedge safety shown in Fig. 4, the spring S must be just tight enough to prevent the wedge W being pulled upwards when the car is descending by the guide rail A coming in contact with it. A weakness of the spring S frequently causes wedges to rattle. The wedge should move perfectly free and should be frequently examined to see that it does. If, when the safety wedges move freely and the springs S are sufficiently tight, the wedges are still thrown into action or rattle when the car descends, the probability is that one of the cables has stretched or is broken. Care must be taken that all cables draw alike; when they do, the equalizing lever G should be horizontal, as shown in Fig. 4 (*a*). In this position the setscrews s, s' should not touch the finger H, but should be so adjusted as to touch and move the finger when the lever G is tipped a certain amount either way. The governor should not be too sensitive to harmless variations in car speed. For this reason, the governor rope r acts on the lever l through the intermediary of a spring, as shown in Fig. 3. This spring should be just tight enough to prevent the wedges from rattling when the car is moving at its normal speed, but not tighter, or the usefulness of the governor will be destroyed.

16. Guides should not be allowed to become gummy, for in this condition they are apt to cause much trouble; they frequently cause the safety wedges to stick, to be thrown into action unnecessarily, or, at least, to rattle. The governor should be examined frequently.

17. In case the safety has acted and has stopped the car, it is of the greatest importance to see, before unlocking the safety, that there is no slack in the hoisting cable. If there is slack, carefully take it up very slowly, reversing the motion of the motor and running it slowly. In hydraulic elevators, this can be done generally by carefully opening the controlling valve; in electric elevators, it is better to turn the worm-shaft by hand. After the slack has been taken up, unlock the safety catches. Most safeties are so arranged that they unlock automatically when the car is moved upwards. Thus, in the Otis gravity-wedge safety the wedges will drop back by gravity. In the safety shown in Fig. 5, the grip roller will readjust itself. In the safety shown in Fig. 7, the governor rope will automatically release itself when the car is going up, but the tripping device must be readjusted by hand. A hole in the car floor is provided for that purpose.

In case the car has been stopped above the top landing, it may become necessary to remove the limit-stop button on the shipper rope, so that the car may be raised high enough to unlock it. If this should prove insufficient, it may even become necessary to raise the car by a tackle.

AIR CUSHIONS.

18. The car safeties treated in Arts. **2** to **17** are designed to act immediately after the slacking or the breaking of a cable, or at the attainment of an excessive car speed. If, when the cable breaks, the car safety should fail to work, owing to neglect or some other cause, the car will drop unretarded to the bottom of the hoistway, causing destruction of property and the probable death of the passengers. An always-ready means of preventing such serious accidents is the **air cushion**. This may be formed by extending the hoistway below the lowest landing in the form of a pit, which has a cross-section at its top somewhat larger than the platform of the car and which gradually tapers towards the bottom to nearly the same cross-section as the platform. When the car falls into this pit, the air within it is compressed and is forced out gradually around the platform of the car, thus letting the car down gradually.

19. Air-cushion pits, in order to be effective, should have a depth equal to one-fifth the whole lift of the car, that is, 20 feet for each 100 feet of hoistway. The walls of the pit must be air-tight, and great care must be used in their construction. Owing to local conditions, it is not always possible to extend the pit far enough below the ground to make it efficient, in which case it may be formed by making the lower part of the hoistway air-tight, say for one or two stories, and providing it with air-tight doors. The engineer in charge of the plant can only see that the pit is not filled with rubbish and when there any doors that they close air-tight.

ACCESSORIES.

SAFETY APPLIANCES.

ELEVATOR ENCLOSURES.

20. The question of **elevator enclosures** is largely a matter of city ordinances. In general, it may be said that every possible means should be taken to prevent accident to passengers on the elevator, as well as persons whose duty brings them near elevator shafts and hatchways. Whatever means are taken by the builders, either of their own account or in compliance with city ordinances, it is the duty of the engineer in charge to see to it that all enclosures are kept in proper condition. He should be constantly on the lookout for improvements in this line.

Whenever possible, elevator enclosures should extend from floor to ceiling, to prevent anything that is being carried on the car catching between its platform and the ceiling. No projections whatever should extend into the hoistway. If full enclosures are not practicable and goods are carried that are liable to stick out, such as rods and similar articles, a car should be used that is enclosed on at least three sides.

Full enclosures need not necessarily be solid walls or partitions, but can be made of lattice, or grille, work substantially braced. As a matter of fact, solid walls for elevator shafts, while recommended by some engineers, are of doubtful value. An elevator shaft so constructed will act, in case of fire, as a chimney, and will carry the flames from one floor to another. Besides, such shafts are apt to be dark unless windows are arranged in them, which make the shaft more dangerous in case of fire. The windows in such shafts should be securely fastened and preferably covered with wire screens. Latticework enclosures will admit plenty of light. In case enclosures are not carried up to the ceiling, they should be at least 5 feet high. Many an accident has occurred by people bending over too low enclosures to look for the car, which then struck them while coming down. Passenger-elevator enclosures are usually made of artistically formed wrought iron and are intended as an ornament to the building in addition to their usefulness. They are generally expensively varnished and should, therefore, be treated with care. They should be cleaned with a feather duster and soft rags. The use of gritty substances, soap, or oil should be avoided. They should be revarnished from time to time, especially after repairs have been made.

ELEVATOR DOORS.

21. Requirements.—**Elevator doors** should always be, if possible, sliding doors or gates so hung that they will operate very freely. They should be provided with latches or locks that can be opened only from the *inside* of the shaft, but they should open easily; that is, without requiring much exertion on the part of the operator. Self-closing doors are to be preferred. The operator should not, however, rely on these self-closing devices, but should always make sure that the door is closed before he leaves the landing with his car. He will and should be held strictly responsible for accidents due to doors having been left open.

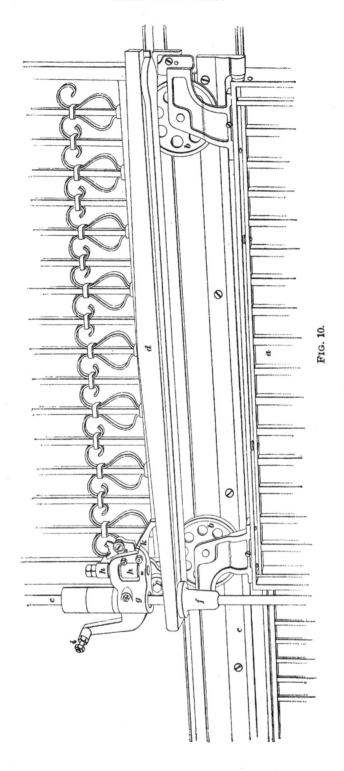

Fig. 10.

22. Self-Opening and Self-Closing Elevator Doors. Various devices are used by different manufacturers to make an elevator door self-opening and self-closing. These devices, in general, have for their object the automatic closing and locking of the door immediately upon the elevator car leaving a landing, and, in addition, are so designed that the operator can open or close the door at will without touching it while the car is at one of its landings and at rest.

23. The elevator-door operating device made by the Winslow Brothers Company, Chicago, Illinois, is shown in Fig. 10. The operation of this device is purely mechanical, the door being moved either way by a friction cone engaging either side of a suitable bar rigidly connected to the door. The construction of the device is as follows: The door a is supported by rollers b, b upon a level track c having a **V** groove planed in it to receive the **V**-shaped rollers. This arrangement prevents any side motion of the door. The so-called **traction plane** d is rigidly attached to the two door hangers that carry the rollers. A vertical shaft e carrying a friction cone f and also a cone-operating device at the top of each landing extends from the top to the bottom of the elevator shaft and is continually revolved by a small electric motor, or from some other source of power by belting. The so-called swing bar g is pivoted to a bracket h that is rigidly fastened to the transom above the door; the swing bar carries a bushing so fitted as to allow it to swing a little. The revolving shaft e, which owing to its length is quite flexible, passes through the bushing of the swing bar, the said bushing forming a journal for the shaft. The free end of the swing bar carries the adjustable buffer i intended to come in contact with a vertical shoe placed on top of the car. This vertical shoe can be thrown forwards so as to press against the buffer, and hence can be made to swing the swing bar around its pivot by a treadle in the car operated by the foot of the operator.

The traction plane is slotted, the slot being beveled and wider at the bottom; by pressing the buffer i away from the car the friction cone will be pressed against the side of the slot nearest the transom and the revolving cone will thus open the door.

As soon as the operator removes his foot from the treadle, the shoe on the top of the car will move away from the buffer i and the shaft will spring back, bringing the friction cone against that side of the slot in the traction plane that is farthest from the transom; the revolving cone will then, by its friction against the surface with which it engages, cause the door to close.

As has just been explained, the door closes whenever the shoe on the top of the car is moved out of contact with the buffer i. This shoe is quite short; consequently, should the operator forget to remove his foot from the treadle in the car when starting the elevator, the movement of the car will very quickly take the shoe vertically out of engagement with the buffer i; the revolving shaft e will then immediately spring back to its normal position and the door will be closed automatically.

The door is held open automatically while the car is at a landing by virtue of a recess in the end of the traction plane into which the friction cone passes after opening the door. The door after closing is locked automatically by a catch k.

24. The Burdett-Rowntree Manufacturing Company use a horizontal pneumatic ram at each landing to automatically open and close the door. The piston of the ram is attached by a link to a long swinging lever connected to the door, and as the ram piston moves one way or the other it carries the door with it. The device is so designed that the door is always held closed until the car is at a landing, when the operator, by pressing on a treadle, throws a movable vertical shoe against a suitable part of the valve gear. This operation unlocks the door and admits air under pressure to one side of the piston in the ram cylinder, at the same time opening the other side to the exhaust. The door now opens, and when wide open can be kept so by a finger lock as long as the car is at rest. Whenever the operator removes his foot from the treadle, or unlocks the finger lock, or starts the car either way without having closed the door, the door closes automatically by reason of the valve gear operated by the shoe on the car returning immediately to its normal position.

25. Car-Locking Device.—With elevator doors that are operated directly by hand by the operator, a **car-locking** device is sometimes used that automatically holds the car in position at its landings and only releases the car when the door is fully closed. While such devices are called car-locking devices, it must not be inferred that they lock the car itself to the landings or to the guides; instead they lock the operating device in the car so that the operator cannot move it to start the car in case the door has been left open.

26. Fig. 11 shows the car-locking device designed by Messrs. I. S. Muckle and W. H. B. Teamer. In Fig. 11 (a), the car A is shown at one of its landings and at rest, in which position the operating device F occupies its central position. The door D is unlatched and opened, as shown in Fig. 11 (b) and (c); the operating device in the car is then locked.

The following description of the device is partially taken from the patent specifications: Secured to one of the floor-beams within the elevator well is a spring latch E, which is bent as shown in Fig. 11 (a), and extends up into the path of an arm d secured to the door D. This arm is notched at d_1 to receive the spring latch E when the door is closed. When the latch is in the notch of the arm of the door, the latter cannot be moved until the latch is pushed out of the notch by the mechanism carried by the car; the door will then be free to be opened.

A pinion f_2 is keyed to the shaft f_1 of the operating device in the car and meshes with a gear f_3 turning on the stud f_4. A crankpin on this gear f_3 is connected by a rod f_5 to the lever f_6 pivoted at f_7 to a bracket a_1 fastened to the bottom of the car. A bearing a_2 on the bottom of the car carries a

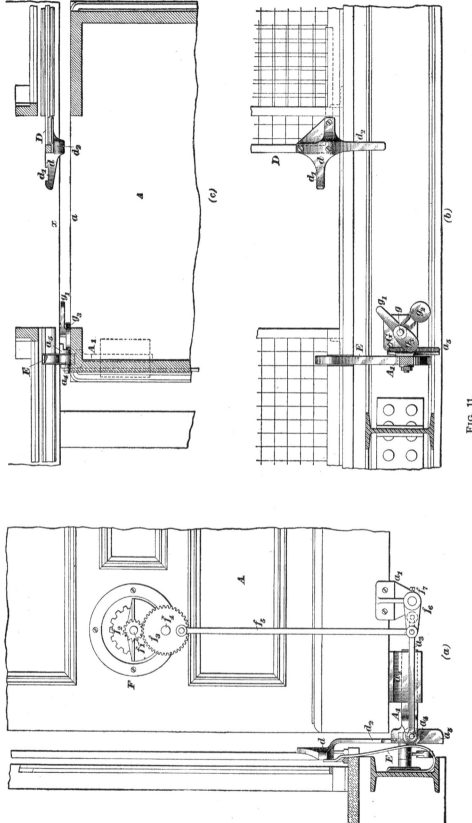

FIG. 11.

slide A_1, and this slide is connected to the lever f_6 by a rod a_3. It is readily seen that, by virtue of the manner in which the parts are connected, the slide A_1 will be in its extreme outer position when the operating device is in its central position, as shown in Fig. 11 (*a*). The slide A_1 carries a roller a_4 that engages with the spring latch E and forces it out of the notch d_1 of the arm d carried by the sliding door, releasing the latter.

It is seen from the above description that the combination of the slide A_1 with the operating device constitutes a mechanism adapted to release the sliding door whenever the operating device is moved to stop the car, that is, is moved to its central position.

It will now be shown how the operating device is rendered inoperative, i. e., how the operator is prevented from starting the car while the door is open. On the face of the elevator well, to one side of the spring latch E, is a plate G carrying a stud g on which is hung a three-armed lever. The arm g_1 of this lever extends in the path of an arm d_2 depending from the door, so that the opening of the door allows the lever, under the influence of the weight g_2, to turn to the position shown in Fig. 11 (*b*). In this position the arm g_3 of the lever has passed behind a flange a_5 of the slide A_1 and prevents the slide from being drawn towards the car. Consequently, the operator cannot move his operating device to start the car, since this can only be moved when the slide A_1 is free. On closing the door, the dependent arm d_2 of the door engages the arm g_1 and turning the lever about its fulcrum g moves the arm g_3 out of the way of the flange on A_1, thus unlocking the slide and hence the operating device.

TRAP DOORS.

27. In many instances it is impractical to erect enclosures of any kind, as, for instance, when the elevator is located in the center of a warehouse and must be accessible from all sides. In such a case, the holes in the floors through which the car passes must be kept covered and must be uncovered only to let the car pass. This is best done automatically in some such manner as is shown in Fig. 12. The car is provided

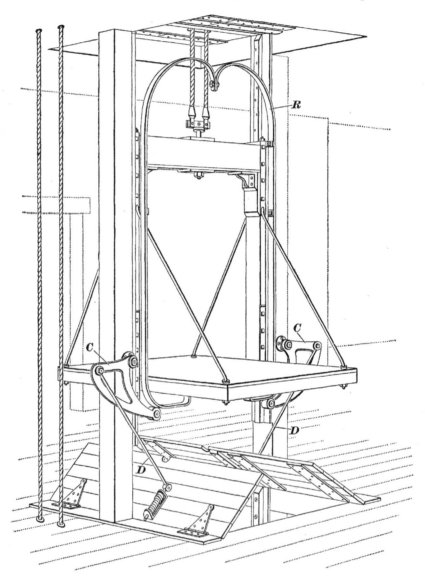

Fig. 12.

with an iron rail R. The arch-shaped upper part of this rail gradually opens the trap doors when the car ascends, and the curvature of the under part lets them down gently when the car descends. To open the trap doors when the car descends, the rail R strikes with its lower portion bell-cranks C, C that are suitably connected to the door by rods D, D.

INDICATORS AND SIGNALS.

INTRODUCTION.

28. Signals must be considered in many cases as a necessary element of safety, especially in freight elevators with insufficient enclosures or trap-door elevators. Electric bells, one on each floor, so arranged that they commence and continue to ring while the elevator passes the floor, are excellent safeguards; they not only warn persons against the approaching car, but tend towards the prevention of any attempt being made to operate the elevator from two floors at the same time.

29. For passenger service, a signal is necessary to communicate with the operator in the car from each floor. This is done very simply by means of a so-called **annunciator** placed in the car and a push button on each floor near the elevator door. Where the traffic is but slight, this means of communication is satisfactory enough; but where the service is rapid, it proves insufficient. Generally in such cases there are, at least, two elevators running all the time, one going up, the other down, and the would-be passenger should know which one to signal. For this purpose, so-called **indicators** have been devised, which show on each floor simultaneously the whereabouts of the car and whether it is going up or down.

MECHANICAL INDICATOR.

30. A simple mechanical device of this kind is shown in Fig. 13. On the shaft A of the overhead sheave is mounted a worm D meshing with a worm-wheel E that is mounted on a shaft F. This shaft carries a chain wheel I, from which motion is transferred by a chain *N* and rods *T* down the elevator shaft to each floor. The rods *T* are guided in plates *W*, one on each floor, and carry arms *Z*, *Z*. From these arms cords are carried over idlers *X*, *X* mounted on the

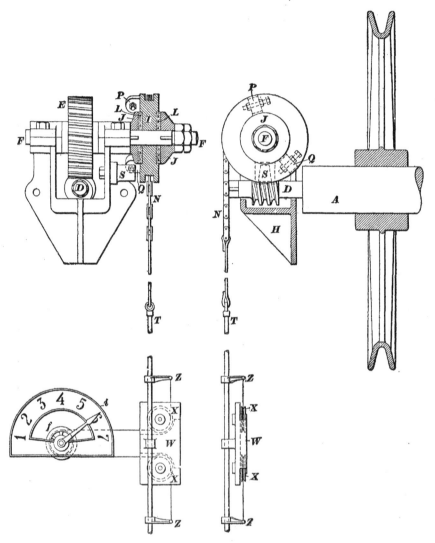

FIG. 13.

plates *W* and around small sheaves *f* in dial plates *i* attached at conspicuous places near the elevator doors. It will be understood that as the car travels up or down, the dial hand will move over the figures displayed on the dial and thus indicate the position of the car. The apparatus is made self-adjusting to rectify any disarrangement due to slipping of the chain.

The wheel I only makes a part of a revolution. It is provided with lugs P and Q that strike a stop S fixed to the frame of the machine as the car reaches its uppermost or lowermost positions, respectively. In case the apparatus has become deranged and indicates wrong, the one or the other of the lugs P, Q will strike the stop S before the car reaches its extreme point of travel and will bring the chain wheel I to a stop. On the return trip, the apparatus will then be readjusted. The chain wheel proper is mounted loosely on its shaft F and is clamped thereto by friction disks J, J fast to the shaft and leather washers L, L.

ELECTRIC SIGNALS AND INDICATORS.

31. The enormous traffic that has to be handled in the large office buildings has called for still more elaborate means of signalling than those afforded by annunciators and indicator dials. In such buildings the service is practically continuous and very swift; the operator has no time to consult an annunciator to find out on which floor passengers are waiting. On the other hand, a passenger standing in front of a row of swift-running elevators and wishing to get the next car would have, if he were to consult indicator dials, to patrol up and down in front of the elevator doors, and would be likely to miss several cars running in the direction in which he wants to go.

32. The usual plan followed in such cases is to provide a signal which, when operated by the passenger, will be noticed by the operator on every car of the series early enough for him to stop at the particular floor where the signal was given. The first car conductor answering the signal then destroys all the signals in the other cars. This plan has been successfully carried out in the Armstrong system, handled by the Elevator Supply and Repair Company, of

New York. This system operates as follows: There are several push-button plates of two buttons, the one marked *up* and the other *down*, conveniently located on each floor. Over each elevator door is a double-light electric lantern, one light marked *up* and the other *down*. A passenger desiring to signal the first car of a bank of elevators, pushes either the "up" or "down" button. This sets the signal, and when the first car moving in the direction the passenger wishes to go reaches a point about three floors distant from that on which he is standing, the lamp in the "up" or "down" compartment of the signal lantern on the outside of the elevator enclosure is automatically illuminated. When the first car approaching the waiting passenger going in the direction he wishes, either up or down, reaches a point about one floor distant, the "operator's signal" is flashed, giving him ample time to stop his car before running past the floor. The operator's signal is a small lamp inside the car constantly in sight. The lamps in both the lantern and car fixture remain illuminated until the car has left the floor from which the signal was given.

There can be no confusion of signals, because the operator can never have but one signal at a time. Moreover, the system is entirely automatic. It allows the operator the free use of his hands and he can thus give all his attention to the control of the car and the safety of his passengers. When no signal light appears in the car, the operator can run at full speed, knowing that no passengers are waiting. Should the first car that receives the signal be fully loaded and therefore unable to stop for more passengers, the operator may transfer the signal to the next car by pushing a button.

All this is accomplished by means of so-called commutators, one for each elevator, placed at the top of the shaft and run by a belt or chain from a pulley on the overhead sheave shaft, in connection with a number of electromagnets corresponding to the number of floors in the building. We forego a detailed description of the apparatus and the electrical connections thereof, since once installed, the apparatus

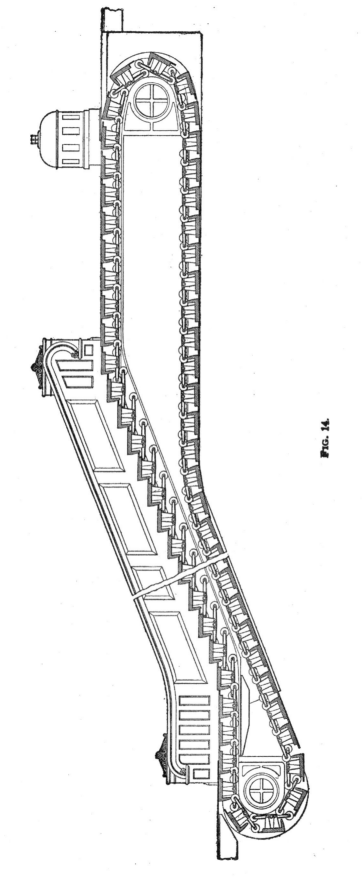

Fig. 14.

needs never to be disturbed. The engineer in charge should see to it that the contacts are kept clean and that the mercury cups used to make the various circuits have the proper amount of mercury. The current for the push-button circuits is furnished by a small motor-dynamo transforming an ordinary 110-volt lighting circuit to one of about 10 volts. This motor-dynamo, of course, needs an occasional inspection, just the same as the other machinery. The current for the lanterns is taken from the lighting circuit direct.

ESCALATORS.

33. The name **escalators** has of late appeared in the terminology of elevator practice for what are commonly known as **moving stairways**. These moving stairways are, properly, not to be classed among elevators, being constructed upon entirely different principles and are mentioned here only for sake of completeness and for the reason that they are destined to take the place of elevators in many instances. Thus it has been found that for short lifts, say one or two stories high, and where great numbers of people are to be transported, that adequate elevator capacity can be had only at great expense and sacrifice of floor space out of keeping with the profits accruing therefrom.

The moving stairway consists of an endless chain, to which are attached steps in such a manner that they form steps like those of an ordinary stairway. By an arrangement of cams, guide rails, and rollers these steps form a plane surface at the bottom and top landing. The accompanying sketch, Fig. 14, will convey the idea. It represents one of the latest designs of this class of passenger-transportation machinery built by the Otis Elevator Company, New York.

CPSIA information can be obtained
at www.ICGtesting.com
Printed in the USA
BVHW05s0328051018
529195BV00005B/240/P